Valuing Health Risks, Costs, and Benefits for Environmental Decision Making

REPORT OF A CONFERENCE

P. Brett Hammond and Rob Coppock, Editors

Steering Committee on Valuing Health Risks, Costs,
and Benefits for Environmental Decisions

Board on Environmental Studies and Toxicology
Commission on Physical Sciences, Mathematics and Resources
Commission on Behavioral and Social Sciences and Education
National Research Council

NATIONAL ACADEMY PRESS
Washington, D.C. 1990

NOTICE: The project that is the subject of this report was approved by the Governing Board of the National Research Council, whose members are drawn from the councils of the National Academy of Sciences, the National Academy of Engineering, and the Institute of Medicine. The members of the committee responsible for the report were chosen for their special competences and with regard for appropriate balance.

This report has been reviewed by a group other than the authors according to procedures approved by a Report Review Committee consisting of members of the National Academy of Sciences, the National Academy of Engineering, and the Institute of Medicine.

The National Academy of Sciences is a private, nonprofit, self-perpetuating society of distinguished scholars engaged in scientific and engineering research, dedicated to the furtherance of science and technology and to their use for the general welfare. Upon the authority of the charter granted to it by the Congress in 1863, the Academy has a mandate that requires it to advise the federal government on scientific and technical matters. Dr. Frank Press is president of the National Academy of Sciences.

The National Academy of Engineering was established in 1964, under the charter of the National Academy of Sciences, as a parallel organization of outstanding engineers. It is autonomous in its administration and in the selection of its members, sharing with the National Academy of Sciences the responsibility for advising the federal government. The National Academy of Engineering also sponsors engineering programs aimed at meeting national needs, encourages education and research, and recognizes the superior achievements of engineers. Dr. Robert M. White is president of the National Academy of Engineering.

The Institute of Medicine was established in 1970 by the National Academy of Sciences to secure the services of eminent members of appropriate professions in the examination of policy matters pertaining to the health of the public. The Institute acts under the responsibility given to the National Academy of Sciences by its congressional charter to be an adviser to the federal government and, upon its own initiative, to identify issues of medical care, research, and education. Dr. Samuel O. Thier is president of the Institute of Medicine.

The National Research Council was organized by the National Academy of Sciences in 1916 to associate the broad community of science and technology with the Academy's purposes of furthering knowledge and advising the federal government. Functioning in accordance with general policies determined by the Academy, the Council has become the principal operating agency of both the National Academy of Sciences and the National Academy of Engineering in providing services to the government, the public, and the scientific and engineering communities. The Council is administered jointly by both Academies and the Institute of Medicine. Dr. Frank Press and Dr. Robert M. White are chairman and vice chairman, respectively, of the National Research Council.

L.C. card no. 89-64210
ISBN 0-309-04195-3

Available from:
National Academy Press
2101 Constitution Avenue, N.W.
Washington, D.C. 20418
S096
Printed in the United States of America

Contents

Preface

A major analytical and policy challenge facing government officials is how to evaluate the risks, costs, and benefits of health and safety policies. This volume contains essays that address philosophical, political, and economic aspects of evaluating programs that ameliorate risks to life. As such, it is the third in a series of studies about risk policy undertaken by the National Research Council (NRC). In 1983 the NRC published *Risk Assessment in the Federal Government: Managing the Process*, which focused on improving policy-relevant scientific descriptions of risk and risk decisions within the government. More recently, the NRC reported on its examination of measures for improving social and personal choices on technological issues by better risk communication (*Improving Risk Communication*, 1989).

The project on valuing risks to life and health was initiated in response to a request from the Environmental Protection Agency (EPA). The EPA and other regulatory agencies have sought to develop analytically sound and politically feasible approaches to gauging the costs and benefits of programs to reduce various risks associated with national environmental policies. In pursuing this objective, regulatory agencies have encountered differing guidance from the courts, Congress, and the Office of Management and Budget regarding the use of benefit-cost analysis in regulatory decision making. These signals led the EPA to ask the NRC for assistance in identifying some sound scientific basis for approaching the problem of valuing risks. Because many of the different points of view about applying benefit-cost analysis to environmental health and safety regulation reflect differing scientific, economic, philosophical, and administrative assumptions, the NRC seemed to provide an ideal forum for a major discussion to clarify underlying issues and distinctions and to point toward areas in which practical analytical and procedural solutions might be sought.

In response to this request, the National Research Council formed the Steering Committee on Valuing Health Risks, Costs, and Benefits for Environmental Decisions. In order to reflect the breadth of issues to be addressed, the committee was made responsible to two units of the National Research Council: the Commission on Behavioral and Social Sciences and Education and—within the Commission on Physical Sciences, Mathematics and Resources—the Board on Environmental Studies and Toxicology. The committee represents a cross section of relevant experience and expertise and includes experts on the economics, philosophy, and politics of risk and health and safety regulation; the use of scientific risk assessment in valuing risks; and the management of the regulatory process.

The committee's charge was to identify areas of disagreement and agreement regarding the use of benefit-cost analysis for environmental health and safety regulation, as well as ways the government might begin to resolve disagreements. The committee concluded that the participants in the policy process disagree about several fundamental underlying assumptions of benefit-cost analysis. Consequently, the committee could not formulate a "manual" for conducting risk analyses acceptable to all policy participants. It concentrated instead on identifying key issues and procedures that might form a basis for developing common understandings.

In planning the conference, the committee met twice in the fall of 1986 to formulate key policy-relevant economic, political, philosophical, and scientific issues, to commission papers from leading scholars, and to formulate a conference agenda. The conference was held in June 1987 and brought together approximately 100 government policy analysts, policy makers, legal and environmental health experts, academic scholars (e.g., economists, political scientists, philosophers, natural scientists) and journalists for two days to discuss issues raised by commissioned papers. Lively discussions of these issues were grounded both by a case study on fugitive arsenic emissions prepared by EPA staff (see the Appendix) and by the comments of agency analysts, jurists, congressional staff, and other participants familiar with the practical challenges of environmental, occupational, transportation, and other health and safety issues.

The principal product of the conference is this volume which contains an introduction (Chapter 1) and a summary of conference discussions and conclusions (Chapter 8), prepared by the committee. Between these two chapters there are six individually authored papers representing the views of the scholars commissioned by the committee to stimulate discussion on key issues. The views expressed in these papers are those of the authors and do not necessarily reflect positions taken by the committee.

The committee wishes to acknowledge the contributions of several individuals and organizations to the conference and to the report. In particular, the Environmental Protection Agency, especially the EPA's Office of Policy Analysis, provided support for the conference, and its director, Richard Morgenstern, and analytical staff members, Robert Wolcott and Mark Thayer, all participated fully in the conference.

Although the report's introduction and conclusions represent the views of the committee, it would not have been produced without the support of the professional staff from the National Research Council, who drafted these chapters and worked with authors in revising their papers: P. Brett Hammond and Rob Coppock. Their intellectual contributions greatly advanced the committee's efforts throughout the project. The report was substantially improved by the diligent work of its editors, Christine McShane and Leah Mazade. In addition, invaluable support was provided by Rose Meadow, Ruth O'Brien, and Carey Gellman.

The committee's conclusions present several insights regarding the need to specify carefully the analytical contributions and limitations of benefit-cost analysis to the problem of valuing health and safety risks. It recommends modest changes in procedures for conducting major analyses, particularly through increased use of peer review mechanisms. The committee's conclusions, if heeded, could enhance the appropriate use of analysis for regulatory policy making.

Roger G. Noll, Chair

Executive Summary

Regulatory agencies regularly confront the difficult task of placing a value on the prevention of death and illness due to encounters (sometimes years earlier) with toxic substances and other health and safety hazards of modern life. Because they require explicit or implicit allocations of scarce resources, regulatory decisions about environmental health attach values to the consequences of those decisions. The problems encountered by benefit-cost analysts responsible for valuing environmental, health, and safety risks are closely related to the legal and administrative context in which they appear, the types of potential threats they pose to life and health, and the characteristics and availability of information about those threats.

With support from the Environmental Protection Agency, a steering committee of the National Research Council planned and conducted a conference addressing these issues. On the basis of its conference discussions and papers, the steering committee identified the challenge facing health and safety regulators in improving the application of benefit-cost analysis to regulation: it is to design practical procedures and techniques that accommodate (1) considerable situational variation; (2) the fairly limited role played by formal benefit-cost analysis in the full process of identifying, regulating, and enforcing solutions to environmental, health, and safety problems; and (3) the tendency for both critics and supporters of analysis to overemphasize its influence in the regulatory process.

Both those who would generally support the use of benefit-cost analysis and those who would oppose its current use recognize genuine moral and ethical dilemmas underlying evaluation of the costs and benefits of programs to regulate health and safety risks. They raise serious questions regarding whether current approaches to characterizing and valuing risks can accommodate the full range of factors that decision makers are asked to take into account, particularly those drawing comparisons across time. They also express concerns regarding the appropriateness of formalizing approaches to issues such as intertemporal equity.

Moreover, although many analysts would agree on the use of certain specific benefit-cost approaches and techniques—for example, the use

1

of willingness-to-pay approaches to valuing prevention of deaths—legal statutes, court interpretations, and other forms of policy guidance provide inconsistent signals to analysts regarding the use of general approaches and specific methods for attaching values to death prevention and life extension. There are both debate among researchers and practicing analysts as to the appropriateness of many specific techniques in given circumstances and a puzzling array of situations confronting regulatory decisions makers.

The following recommendations for improving benefit-cost analysis draw on such considerations:

1. Benefit-cost analysis should be thought of as a set of information-gathering and organizing tools that can be used to support decision making rather than as a decision-making mechanism itself.

2. Analytic methods and techniques should be more systematically matched to types of health and safety problems in the regulatory process.

3. Regulatory agencies should consider expanding the use of formal peer review mechanisms in the area of benefit-cost analysis for health and safety decisions.

1
Introduction

Regulatory agencies regularly confront difficult trade-offs among important values, such as health, safety, longevity, and the monetary and nonmonetary costs and benefits associated with protecting human health and well-being against the hazards of modern life. Those responsible for reducing deaths and illnesses due to encounters (sometimes years earlier) with toxic substances and other potential health and safety hazards must often assess uncertain events. They must also balance benefits and costs that are difficult to characterize or measure. Individual analysts and entire regulatory agencies, such as the Environmental Protection Agency (EPA), have devoted considerable effort to improving benefit-cost analysis, cost-effectiveness analysis, and other ways of characterizing trade-offs among key values. However, some analytic approaches and techniques and the details of their application remain controversial.

For example, once scientists have identified the risks involved in problems such as pollutants in the air and water, unsafe factory conditions, dangerous consumer products, and threats from toxic chemicals, regulatory attention shifts to the question of what kind and how many resources should be devoted to reducing or avoiding the risk in question. Consider some common examples: Which policies will do the most to improve the safety of children's toys? Who will pay for the health effects of asbestos exposure? Should we require copper smelters to reduce or eliminate fugitive arsenic emissions? How do we know whether preventing further ozone depletion is worth a particular effort? *Such choices, because they involve inherently scarce resources, attach values to the consequences of the actions involved.*

This volume reports on the results of a conference planned and conducted by an independent committee of volunteers appointed by the National Research Council to address an important set of issues encountered in using formal approaches to valuing risks, costs, and benefits in health and safety policy making. During the 1980s, interest in comparing benefits and costs of proposed regulations was stimulated in EPA and several other

3

federal regulatory agencies, partly in response to a sharper focus by the President and the Office of Management and Budget (OMB) on reducing private-sector expenditures associated with regulation. The difficulty of developing explicit values for human lives extended, disabilities avoided, and others has led these agencies to search for practical yet theoretically defensible approaches to risk valuation and to benefit-cost or cost-effectiveness analysis. As a part of that effort, the committee organized a conference that surveyed recent scholarship in law, philosophy, political science, and economics and discussed what research, if any, would be likely to lead to improved regulatory decision making, particularly benefit-cost analysis.

Neither the conference nor the more general search for improved benefit-cost analysis has produced complete agreement within agencies, across agencies, in Congress, in OMB, and among interest groups about valid ways to improve valuation of risks, costs, or benefits. Rather, the search has revealed, and in some instances sharpened, disagreements about the appropriateness of specific attempts to value risks for health, safety, and environmental regulation and about methods and approaches for doing so. Instead of integrating analysis of environmental, health, and safety risks into complete discussions of regulatory decisions, those who oppose or support particular regulations have often chosen to separate them from consideration of other factors, to politicize the analysis, or both. And disagreement about some of the ethical issues underlying benefit-cost analysis can allow such politicization to continue. Consequently, agencies producing analyses find themselves repeatedly challenged in court and in Congress for exceeding their statutory charters, failing to account for important noneconomic concerns—such as the special nature of human life—or inadequately addressing other issues, such as effects that would be felt only a very long time in the future.

The central policy-making problem for health and safety regulation is thus the very practical question that assisted the committee in organizing the conference, as well as its conclusions and recommendations: What is gained by explicit treatment of the inherent values and trade-offs? Since any regulatory decision involves resource expenditures by individuals, firms, government, or other parts of society, some examination of resource expenditures and what is received for them seems appropriate. But how should this be done, especially when intangibles like human health are involved? Is there any prospect for improvements in benefit-cost analysis that would produce increased consensus on its underlying assumptions and application? Interest has grown among regulators, scholars, and analysts in exploring some of the most troublesome issues underlying valuation techniques and approaches in an effort to understand the sources of disagreement and the possibility for achieving more consistent, valid, and practical approaches. These issues include:

(1) administrative, legal, and statutory constraints on application of benefit-cost analysis techniques;
(2) the philosophical foundations of using a common monetary metric to represent human life and health;
(3) moral issues involved in comparing health-risk benefits and costs across time and distance, such as individual and collective preferences for who should bear the burdens or benefits of risks across time and distance;
(4) the adequacy of scientific information as a basis for placing values on health risks in particular instances; and
(5) effects on regulatory policy of new research knowledge about decision making.

The balance of this introductory chapter further explores questions raised for researchers, regulatory policy makers, and agency analysts by the use of the valuation techniques we have labeled benefit-cost analysis.

RISK ASSESSMENT AND BENEFIT-COST ANALYSIS

In addressing environmental, health, and safety problems, regulatory agencies are often asked to develop an understanding of both risks and resources. They are asked to assess the risks associated with specific health and safety hazards. They may also analyze the costs and benefits of alternative ways of reducing those risks. Although they operate under different statutes (in some cases, several for one agency), EPA and other regulators—such as the Occupational Safety and Health Administration, the Food and Drug Administration, and the Consumer Product Safety Commission—have been required for more than a decade to formally assess the health risks associated with exposure to toxic substances and other potential hazards, including air and water pollution, food additives and contaminants, and products and processes in the workplace. During that time there have been a number of efforts to improve the physical, chemical, and biological bases for policy making. More recently, there have been several efforts to examine the process for assessing risks at the Environmental Protection Agency and other federal health and safety regulators.

One such effort, by a committee of the National Research Council, examined the process by which federal agencies characterize and determine the risks of cancer for policy-making purposes. The committee's report, *Risk Assessment in the Federal Government* (National Research Council, 1983), examined procedures used by agencies in the early 1980s to determine cancer risks. It proposed that development of scientific information about specific hazardous risks be separated analytically from decision-making and management activities of a regulatory agency. Such a separation would

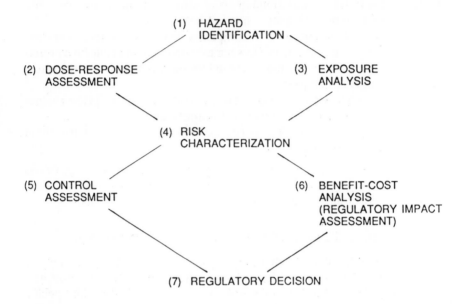

Risk Assessment

(1) HAZARD
     IDENTIFICATION

(2) DOSE-RESPONSE   (3) EXPOSURE
     ASSESSMENT   ANALYSIS

         (4) RISK
              CHARACTERIZATION

(5) CONTROL   (6) BENEFIT-COST
     ASSESSMENT   ANALYSIS
                                                 (REGULATORY IMPACT
                                                 ASSESSMENT)

         (7) REGULATORY DECISION

Risk Management

FIGURE 1   Risk assessment and management process.

allow for scientific criteria to be the primary criteria in guiding research and data gathering.

The report also constructed a framework for organizing diverse kinds of hazard-related information needed for a complete program of risk management; it is summarized and adapted in Figure 1. In this scheme an agency might begin with a careful process of risk assessment, including data gathering, scientific testing, and evaluation of results, to learn more about human and animal responses to various doses of a potential hazard and about what exposures to that hazard people may be experiencing.

The 1983 report acknowledged the reality of contemporary risk assessment by identifying areas in which gaps in scientific data and theory often appear. It proposed that agencies develop guidelines for using conservative, scientifically appropriate inferences to bridge the gaps for policy-making purposes rather than either taking ad hoc inferential leaps or postponing action until all scientific uncertainty is removed. To properly identify the strengths and weaknesses of any assessment, the report emphasized the need for assessers to make assumptions explicit as well as the need for peer reviews of the scientific process and results of assessments. In response to proposals such as these, federal regulators have invested considerable effort in recent years to improving assessments of hazards in order to better characterize the risks involved.

Although the 1983 report focused primarily on risk assessment, several of its conclusions and recommendations are also relevant to the analysis of benefits and costs in support of risk management. The term *benefit-cost analysis* refers to a family of approaches and techniques for expressing alternative ways of achieving an objective in quantitative comparable terms. For example, the classical approach is to express all benefits and costs associated with each alternative in dollars as a function of time, discount the future benefits and costs at some appropriate rate, and then compare the alternatives on the basis of the current value of the net benefits or on the basis of an internal rate of return (Quade, 1982:58, uses this common definition; among many others, see Kneese, 1989:295-298).

This classical approach most closely resembles an ideal definition that cannot be applied in many instances due to practical limitations. More often, agencies use a version of benefit-cost analysis that focuses on the cost-effectiveness of alternative policies. That is, a single benefit (e.g., a reduction in pollution) might be analyzed against several potential options with various costs. The distinction is important in that costs but not benefits are being compared. Nevertheless, this volume focuses on philosophical and other assumptions underlying comparisons inherent in analysis of costs, benefits, or both. Therefore, following common practice, in this volume except as otherwise specified, the term *benefit-cost analysis* also refers to cost-effectiveness analysis and other variations on this theme. For example, after evaluating the potential health risks of a pollutant or toxic substance or consumer product design, regulators may examine the costs of alternative strategies for a uniform reduction in the risk (e.g., eliminate the offending item, mitigate its effects, redesign it, or do nothing), and then compare explicitly the costs of those alternatives. Information about risks, costs, and (when permitted and/or possible) benefits may be summarized and used by regulatory policy makers as part of the basis of a regulatory decision or in order to support one.

REGULATORY PRACTICE

Regulatory practice has been far less consistent than such a scheme would indicate, however. Approaches to valuing health and safety risks have varied considerably by issue, between agencies, even within agencies, and over time. The sources of such differences are numerous. First, with certain exceptions, Congress grants each regulator considerable discretion in carrying out broad legislative mandates. Regulatory agencies are allowed, encouraged, and even required to exercise discretion in implementing statutes. Second, statutes (and case law) vary as well, some specifying that regulators will consider costs and benefits of alternative policies, others actively prohibiting such economic considerations. While most agencies have considered benefit-cost issues in regulatory matters, statutory variation allows some agencies to pursue the latest techniques and approaches more vigorously than others. Third, and possibly most relevant to the current attention being given to regulatory risk valuation issues, is the evolution of presidential interest in and ability to affect the application of cost calculations in regulatory policy making. Efforts by OMB to guide agency use of statutory discretion to promote greater analytic rigor and consistency in federal approaches to regulatory risk valuation have sparked serious debates about the prospects for reaching those goals.

The use of benefit-cost analysis in the federal government is long-standing and well documented. Benefit-cost analysis has antecedents in the early 1930s, as a way of evaluating public water projects. It emerged on a large scale in the early 1960s in the Department of Defense and in federal recreation programs with the use of the Planning, Programming, Budgeting System (PPBS) and its successors. Application of benefit-cost techniques in health and safety matters grew from several White House policy review mechanisms. As an antecedent to formal benefit-cost analysis, in 1971 OMB ordered EPA to submit significant regulatory proposals to an interagency committee for nonbinding discussion. Subsequently, President Ford's 1974 executive order required agency heads appointed by the President to submit for OMB review Inflation Impact Statements (later entitled Economic Impact Statements) on proposed regulations likely to have substantial economic effects. This requirement was used to prod agencies to explain more explicitly the likely economic effects of regulations, but no consistent agreement emerged as to how these statements would or would not be used to inform regulatory decisions.

A Carter administration executive order issued in 1978 created a new interagency forum, the Regulatory Analysis Review Group (RARG), chaired by the head of the Council of Economic Advisers, to discuss economic analyses (Regulatory Analyses) of 10-20 key proposed regulations each year. Although benefit-cost analysis was not specifically part of the RARG charter, agencies were asked to include "an analysis of the economic consequences of each . . . alternative." In addition, agencies were directed to solicit public participation, to choose the least burdensome alternative, and to justify the choice. The RARG then reviewed, discussed, and made public nonbinding comments on these analyses and proposed regulations.

Within a month after taking office, the incoming Reagan administration issued Executive Order 12291 (February 17, 1981), which replaced the

RARG process with OMB's formal review and clearance of proposed regulations and which required agencies to prepare and to base all major new regulations on formal benefit-cost analyses. The White House emphasized that, within an agency's statutory constraints, the principles guiding regulatory decisions should be twofold: first, that regulation should be used only in instances in which it is superior to market forces and, second, that regulation should be demonstrated to maximize net benefits to society. More specifically, notable provisions of E.O. 12291 included requirements that regulatory actions should not be undertaken unless potential benefits to society from the regulation outweigh the potential costs to society; among alternative approaches, the alternative involving least net cost to society should be chosen; and that agencies set regulatory priorities to maximize aggregate net benefits, taking into account the condition of particular industries involved. These requirements provided the rationale for OMB's effort to increase the use of formal techniques for determining net benefits to society. In the case of environmental, health, and safety regulation, the executive order prompted OMB to require further efforts on the part of agencies to value alternative policies for prevention of deaths, injuries, and illnesses due to environmental and other hazards.

Overall, during the eight-year Reagan administration, OMB reviewed approximately 30 regulatory impact analyses done in support of major new proposed regulations. Between 1981 and 1985, 15 analyses were submitted by EPA in the areas of environmental health and safety regulation. In addition, during this same period of time, OMB chose not to review dozens of other benefit-cost and cost-effectiveness analyses performed by EPA and other regulatory agencies. OMB may have been limited in its coverage of analyses by the size of its own analytical staff. Of approximately 35-40 professional staff in OMB's Office of Information and Regulatory Affairs, nearly all were devoted to regulatory matters in the years 1981-1989. Nevertheless, the number of regulations issued by EPA alone in that period totaled over 2,000.

CONTINUING ISSUES

In actual regulation of health and safety problems, however, the regulatory agencies' use of formal approaches to valuing risks, costs, and benefits remains unsettled. Particular issues associated with formal valuation remain controversial and the subject of continuing debate. For example, there is no single approach or set of methods for benefit-cost analysis now in use in any federal regulatory agency with responsibility for health and safety. At EPA, for example, of 15 Regulatory Impact Analyses submitted to OMB for review between 1981 and 1985, 6 included direct comparisons of benefits and costs of regulatory alternatives, and 9 concentrated primarily on costs of alternative actions but not benefits (EPA, 1987). To become accepted, even within one agency, any general approach or set of methods would have to address and settle a continuing set of controversies concerning methodological, conceptual, and moral challenges in characterizing and valuing environmental and other health and safety risks. It was this situation that led EPA to request formation of the committee that prepared this conference report.

In his keynote address to the conference (Chapter 2 in this volume), Milton Russell, as former Assistant Administrator for Policy, Planning, and Evaluation at the Environmental Protection Agency, points out graphically the importance of these questions to the day-to-day decisions of regulatory agencies and just how difficult it is to resolve the issues they raise. To illustrate the point, he asks the question "Where should the sludge from municipal sewage plants be disposed?" By pointing out that every option includes at least some risk to human health and reminding us that the sludge has to go somewhere, he demonstrates how difficult it is to reach satisfactory answers. Despite the difficulties and the imponderables, Russell takes the position that it is irresponsible for a public official to decide without tracing the consequences to the extent possible. In real life, public officials face messy, incomplete answers. Russell feels that the legitimacy of their decisions has to derive from the acceptance of those decisions by the people affected, and that the provision of information about the likely consequences is the basis for principled accountability.

Contextual and Legal Constraints

Benefit-cost analysis, like any set of decision tools, is affected by the characteristics of the setting in which it is applied, including the specific statutory requirements and restrictions, the distribution of policy-making responsibilities, and the characteristics and behavior of a multiplicity of interest groups. In order to gauge the possibility for developing more consistent approaches to risk valuation, it is important to understand variations among agencies in terms of such constraints.

- What are the scientific, administrative, and political sources of support for considering economic efficiency as an integral part of risk-management decisions, and what are the constraints on doing so?
- How do statutes, case law, presidential directives, congressional pronouncements, and administrative practice provide guidance in the development and application of benefit-cost analysis for regulatory decisions?
- What are the key variations in the use of risk-control analysis by federal health and safety regulators?

In "The Politics of Benefit-Cost Analysis" (Chapter 3), R. Shep Melnick focuses on the political and administrative forces that block wider implementation of formal benefit-cost analysis for federal regulatory decisions. He outlines the sources of opposition to the use of analysis, as well as the mechanisms that are used by major actors—Congress, public interest groups, OMB, and the agencies—to shape general approaches and specific risk valuation practices. Melnick raises the question of whether the current legal and political context will allow for progress in integrating technical benefit-cost analysis with regulatory decision making on health and safety issues. A lack of progress may be due largely to controversies among key policy makers on the moral validity of issues such as placing a value on human life for regulatory purposes.

Approaches to Analysis

Some analysts and decision makers may find it attractive to value risks, costs, and benefits in terms of a single arithmetically manipulable yardstick that measures all relevant attributes, but various theoretical treatments of value raise problems that must be dealt with in practical applications of the approach. There is, for example, considerable debate about appropriate ways to make comparisons across time and among illnesses, disabilities, and economic or social groups.

- What might be the foundation for a metric covering a variety of health and safety effects?
- Can preference-based economic guidelines be accommodated with political, moral, and other statements, especially with respect to assigning, directly or indirectly, values to human life and health?
- What measurement problems are posed in valuing health risks (e.g., are there appropriate measurement scales, are health effects well ordered)?

In "Benefit-Cost Analysis as a Source of Information About Welfare" (Chapter 4), Peter Railton raises some of the doubts he has as a philosopher about the approach. He addresses these doubts by asking the question "What do I have to believe in order to do [benefit-cost analysis]?" In looking for answers to this question, he considers whether such beliefs seem tenable. For example, he inquires into whether theorists and practitioners of benefit-cost analysis recognize limitations or treat it as sufficient for policy determinations. Railton makes similar inquiries about the assumptions required to believe the scientific measurements on which benefit-cost analyses are necessarily based and about assumptions associated with the blanket application of discounting.

Douglas E. MacLean, in "Comparing Values in Environmental Policies: Moral Issues and Moral Arguments" (Chapter 5), addresses a related set of questions. He reviews briefly several approaches to moral inquiry common in the discipline of philosophy and describes how they might be applied to valuation of health effects in environmental policy decisions. He describes weaknesses in both the "low road" of expressed preferences and the "high road" of theoretical or doctrinal moral reason. Examining central paradoxes in the use of discount rates, especially with regard to the valuation of human life, he points out necessary inconsistencies in the universal application of analytic techniques for making environmental policy decisions and proposes case-specific application of such techniques. In particular, he raises problems applicable in any discussion in which not all costs of a policy are borne by the same individuals or groups as those who reap the benefits. These problems are relevant to intertemporal, interpersonal, geographic, and interillness comparisons of risk.

How Much Information?

It is also important to understand how decision makers cope with information that is imperfect and changing over time, especially since analytic processes and the structuring of data affect the selection of what is included in analyses. This raises a whole series of questions concerning the timing of decisions, their interaction with information-providing activities, and the aims and purposes of the organizations in which they are made.

- Can procedures for valuing health risks be made politically feasible or acceptable; for example, can we agree on mechanisms for deciding who should have a say in determining values and specific regulatory policies?
- How do beliefs about the major consequences of various decisions affect our search for illuminating information?
- What is the relationship between the need for credible valuation and the need for credible decisions?

In "Environmental Policy Making: Act Now or Wait for More Information?" (Chapter 6), Jeffrey E. Harris points out that regulatory decisions about environmental hazards are routinely made in the face of huge uncertainties. He characterizes the central problem in the dynamics of such decisions as that of timing—to act now or hold out for more information. Harris suggests that a principal reason for waiting is the assumption that environmental policies are irreversible. But he observes that in many cases this involves a kind of self-fulfilling prophesy—the longer we wait, the fewer options remain and the more irreversible the decision appears. Harris also suggests that waiting for more research may be equally misguided. The best way to gather information, he claims, may most often be to implement the policies and then to evaluate carefully their consequences. This strategy, of course, assumes that the expected negative consequences of initial implementation are small relative to the consequences of doing nothing.

Handling Uncertainty

Behavioral scientists have begun to challenge the simple theoretical model of rational choice underlying most risk-control analysis, suggesting ways the model might be modified and applied in practical analyses. Much of this work has focused on the handling of uncertainty in formal decision approaches.

- How should analysts and decision makers acknowledge and cope analytically with uncertainties in the science underlying decisions?
- Are we sure enough of our ability to characterize the risks of potential health hazards to use those characterizations as a basis for valuation analysis?
- How do we get better estimates of uncertainty for purposes of practical decision making?

In "Choice Under Uncertainty: Problems Solved and Unsolved" (Chapter 7), Mark J. Machina reviews the expected utility model and describes several problems that beset it, including nonlinearity in probabilities, preference reversal, and framing effects. He examines the empirical evidence regarding each and suggests how these findings have, are likely to, or should change the way economists view and model private and public decisions under uncertainty. On balance, Machina feels the modern models used to characterize behavioral deviation from expected utility theory are completely consistent with well-defined individual preference ordering, and hence with traditional welfare analysis. What to do about framing effects, however, is itself a public policy issue, in Machina's view. We may be able to look to other previous treatments of the presentation of information for guidance.

CONCLUSION

Initial drafts of the papers in this volume served as provocative stimulants for discussion at the conference. Following the papers is a concluding chapter summarizing discussion among conference participants and the steering committee's conclusions regarding the possibility of incorporating the ideas raised into practical approaches to benefit-cost analysis.

REFERENCES

Kneese, Allen V.
  1989   The economics of natural resources. Pp. 281-309 in Michael S. Teitelbaum and
          Jay M. Winter, eds., *Population and Resources in Western Intellectual Traditions.*
          New York: The Population Council. Published as a supplement to Population
          and Development Review, vol. 14. (Also reprinted as Reprint 243 in the RFF
          Reprint Series, Resources for the Future, Washington, D.C.)
National Research Council
  1983   *Risk Assessment in the Federal Government: Managing the Process.* Committee on
          the Institutional Means for Assessment of Risks to Public Health. Washington,
          D.C.: National Academy Press.
Quade, E.S.
  1982   *Analysis for Public Decisions.* New York: Elsevier Publishers
U.S. Environmental Protection Agency
  1987   *E.P.A. Use of Benefit Cost Analysis: 1981-1986.* Office of Policy, Planning
          and Evaluation, EPA-230-05-87-028. Washington, D.C.: U.S. Environmental
          Protection Agency.

2
The Making of Cruel Choices

MILTON RUSSELL

For two days, participants in the National Research Council's conference on valuing health risks, benefits, and costs for environmental decision making considered and debated the ways in which information developed in this process is presented, compared, and evaluated, and how it is to be used or ignored in making environmental decisions.

This debate was an extraordinarily important undertaking. It may be especially useful for those on the "firing line" in agencies such as the Environmental Protection Agency (EPA)—individuals who are in a position to make decisions, to advise those who do, or to prepare the analytic underpinnings that inform the decisions that are made. I know from experience that people in such positions need to set aside time from their day-to-day activities to think critically about the premises underlying their actions. They also need different perspectives, especially the ideas of those who have the opportunity and the inclination to reflect on fundamental issues of environmental decision making. Otherwise, in the press of hour-by-hour activities, they run the risk of relying on rules of thumb and on unexamined value premises of their own—or of others. This conference was designed in part to help those in government, such as myself, carefully consider and critically examine value premises.

My perspective is that of someone who has until very recently been inside the maelstrom—I am not with EPA now—but who has been outside it perhaps long enough to have established some distance. Indeed, this is the third time I have made the journey from academia to government and back. For me, there has been one constant in each of these trips: the way certain operational issues that are so dreadfully important while

Milton Russell is professor of economics and senior fellow at the Energy Environment and Resources Center, University of Tennessee, and collaborating scientist at the Oak Ridge National Laboratory.

in government seem a ludicrous waste of energy a month later, and the increased significance, after more thought, of issues that at first appear to be of less importance. Among the latter are the deeper questions that are at the root of the issue we are examining in this volume: What is the principled basis for government decisions that affect the vital interests of citizens, and from what source do those who make these decisions gain their legitimacy? These questions have engaged some of civilization's best minds for centuries, and I have no illusions that totally satisfactory answers will emerge here. Still, environmental decision making offers a particularly thorny set of issues that must be grappled with, and it is through this grappling that the answers to these questions can be approached.

My experience at EPA revealed that the agency deals with extremely complex problems whose potential solutions have serious and far-reaching implications. I found that an explicit decision framework to sort out pros and cons, benefits and costs, was absolutely essential to any reasonable possibility of using the agency's immense power to do good rather than ill. The implications of decisions were simply too numerous and too diverse to be kept in mind without an explicit mechanism and, to the extent possible, a common metric or standard of measurement to keep score among the trade-offs that had to be made. I also concluded that what was true of the executive branch (to the extent it had discretion under the law) was true as well of the legislative branch as it formulated the statutes. Indeed, as Congress provides more and more detail in the environmental legislation it passes, it faces ever more difficulty in understanding the full implications of its actions—and ever more responsibility to do so.

It is also true, however, that only very seldom does the decision itself leap out of the analysis—that is, unless analysis is broadened so much as to lose its commonsense meaning. For example, the specifics of particular situations, the dictates of protecting a sound decision process, and the implicit signals about what sort of society should be fostered all play a role in producing a decision. A man once remarked to me that the British Navy lost its soul during World War II by issuing the perfectly rational order that convoys were not to stop to pick up survivors of submarine attacks. This policy so violated the tradition of the sea and the honor and respect a great power owed its men as to shake the national resolve. I doubt that this armchair rumination really explains, as this man suggested, the eventual loss of the British Empire, but the point was well taken—decisions that are smart may not be wise.

Nevertheless, the beginning of wisdom in environmental decision making is first to be smart, which, in my view, implies careful, explicit analysis in a structured framework.

Structured in this case does not mean that the elements within such a framework should reflect a static or overly narrow system. For example, an environmental regulation will change the situation to which it applies; consequences follow that need to be considered. Thus, new regulation yields new incentives for technological and managerial improvements that will almost certainly lead to ultimate compliance costs that are lower than those estimated. Or again, the simple fact that processes have to be rethought could overturn established ways of operating and also lead to improvements. The experience of U.S. energy use is relevant here. When the energy price shocks of the 1970s occurred and energy use in this country was examined, most firms and individuals found that they had never bothered to take actions that were well justified even at preshock prices. The same process seems to be under way today in environmental matters and can be seen in the new attention being given to safer disposal and lower production levels of hazardous wastes, the reduced use of pesticides, designs for chemical processes involving the risk of release of toxic chemicals that can handle a broader range of conditions and problems, and so forth. Any analysis of proposed environmental protection actions must take these dynamic effects into account and must also consider the second and higher order effects that follow from the initial perturbation. Often, the result of such consideration is greater risk reduction at substantially lower costs than previously anticipated—although, of course, there may be offsetting problems as well. A careful analytic effort within an explicit framework will help in anticipating these effects.

Another thing I discovered at EPA, though, was that, when it comes to the environment, many people in and out of government are opposed to the use of an explicit framework, especially one cast in benefit-cost terms. There are many reasons for this opposition that I will not detail here. It is my view, however, that one of the most important among them is an almost visceral reaction against the open consideration of any trade-offs regarding human health and the environment, even though such trade-offs are implicit in every decision.

I have some sympathy with this view. It may be that to confront the reality that life has a price, however high, undermines the foundation of a society in which we would want to live. This certainly is the view reflected in the comment on the British Navy convoy policy. It may also be that, when basic values are in conflict, it may seem worthwhile at times to foster the comforting myth of their successful accommodation.

I reject in principle, however, the elitist view that the public cannot be trusted to accept responsibility for cruel choices and that its leaders should instead feed it comforting bromides while making those choices on its behalf. Moreover, to obfuscate inevitable choices is to violate the premise of a government based on the consent of the governed, which to me is the ultimate source of governmental legitimacy in this country.

Besides, there are practical consequences to being less than open and explicit about trade-offs. An explicit analytic decision framework, with quantification to the extent possible, can be critical as a communication device and as a source of discipline for decision makers to prevent their usurpation of power that is not rightly theirs. As a communication device, an explicit decision framework makes obvious at least some of the effects of alternative actions and thereby brings to the surface the bases of decisions. As a result, others may be informed more fully and can make their judgments known. As discipline, it makes it harder for decision makers to hide behind a verbal "fast shuffle" if they seek to impose their own views of the good society on the public without its informed consent.

Analysis of the sort that meets these requirements can take many forms, and EPA uses a rich array of techniques. Analyses range from data-based but ultimately judgmental comparative risk efforts to risk-risk comparisons, cost-effectiveness estimates, and, finally, full-blown, formal benefit-cost studies.

I welcome the discussion of benefit-cost analysis at this conference because of the important issues involved in its use: its value predicates, its unexamined assumptions, its static bias, its demands for data, and, certainly, the opportunity for manipulation of results by unscrupulous practitioners. Yet I hope that the fact that formal benefit-cost analyses have limitations and that their results can be manipulated or overinterpreted does not lead to rejection of the idea that lies behind the motive for using them. That idea proposes that what really counts is to understand what is gained and at what cost from alternative courses of action, and then to make decisions based on the balance that is cast. I have not found a better basis for decisions, or, indeed, in some deeper sense, that there is any other basis— at least when you are in a position in which you *really* must decide what is *actually* to be done.

These are strong statements. In their support, let me offer their predicates as related to environmental protection.

The first predicate is that resources are ultimately limited. There is only so much that can be done, although that amount can be made larger if people work smarter and resources are used more efficiently. The resource "pie" can also expand over time, and resources can be shifted to environmental protection so that the size of that slice of the pie may grow. Yet at any given time, to demand more than exists is an exercise in deluding others; to expect to get it is an exercise in deluding oneself.

The second predicate is that the selection of any action simultaneously rejects others. At the most basic level, labor, materials, and skills devoted to one task cannot be used for another, although it may be impractical to identify the other uses to which the resources would be put.

The third predicate is the corollary of the second. When a regulatory agency rejects one action or technology, it promotes others. For example, to forbid the use of one pesticide promotes the use of alternatives. To prevent sewage sludge from being dumped in the ocean encourages land disposal or incineration.

The final predicate is that decisions *are* made. Ex post facto, there is an array of goods and services produced, health risks that are borne or avoided, and environmental insults that are imposed or turned aside. There are also patterns of individual behavior that are rewarded and those that are penalized, together with social goals that are enshrined and those that are denied.

Given these predicates, the task of public policy is to make the trade-offs that lead to a set of outcomes that are the best possible. And there's the rub: how to decide which are "best."

One view, to which I subscribe and which I think is enshrined in the American system of government, is that what is "best" depends on the values of those to whom government officials are responsible; that is, those now living. This approach does not imply a decision framework that turns its back on the past or one that ignores future generations. Nor does it mean the selfish sacrifice of other elements of planetary life for instant, narrowly human, gratification. Rather, it means leaving those choices to the citizens as a whole, working through established political institutions, instead of allowing some few who feel they know best to arrogate the decision-making role. It is individual citizens who have the responsibility to consider future generations. It is up to them to reflect in their choices the long-term continuity of the natural systems on which they and future generations will depend, and which they cherish. It is up to them to attempt to convince others to adopt their values on these and other matters. Government officials and political leaders have the dual role of first participating in the education and persuasion process and then reflecting in action the goals that are selected.

With respect to whether these goals properly account for the future and for non-human health outcomes, I can only note that decisions are made by humans and that they are being made today. The only issues are which humans, working through what institutions, and reflecting which values. While neither ducks nor those yet unborn may vote, I, along with Jefferson, know of no safe repository of the power to decide other than with those who do.

The U.S. political system, therefore, mediates between the citizens, whose values are to be served, and their agents, whose decisions and actions yield the trade-offs I discussed earlier. This process brings us back to the function of analysis, which is to illuminate the ramifications of choices. It also brings us to the practical necessity of a formal analytic mechanism or set of conventions to collect, organize, summarize, and present information about alternative sets of outcomes to decision makers and the public. The practical questions that follow are how broadly should the net be cast for effects of consequence, what effects caught in that net are relevant, how are they to be valued, and how are they to be presented. As a contribution to the context for a discussion of these issues, I want to provide an illustration of the way some of these questions are presented, and demonstrate why I find an explicit decision framework essential when it comes to protecting the environment.

Municipal sewage plants produce sludge that must be disposed of—on or under the ground, in the air through incineration, or in the ocean. There are irreducible risks in any of these choices.

Land-based disposal carries risks mostly for humans; ocean-based disposal modes primarily affect marine organisms. Some of these risks are incurred immediately: emissions from an incinerator are breathed at the time of disposal, and dietary risk from cropland disposal follows within months. Other risks are incurred in the future: disposal in landfills may lead to the leaching of toxic substances into groundwater, which, even in the event the water were drunk, would not bring exposure for some time. Another distinction among risks is the certainty of the exposure. The air will be breathed, but the water may *not* be drunk. On yet another dimension, emissions from an incinerator may expose a sizable population to risk, although of a very small level, whereas groundwater risks may be greater but would affect only the limited population that someday might draw water from an untested, contaminated well.

In addition, all of these disposal options require resources that could be used to satisfy other needs. The amount spent will vary among options and also within each option with respect to what precautions and controls are imposed. Therefore, the costs to be incurred influence the probability and magnitude of the residual risk.

Furthermore, as noted earlier, any choice that is made and enacted will affect the system as a whole. Dynamic adjustments occur that will often—although not always—yield less perturbation than a static analysis might suggest. Thus, costs are likely to be lower, as are environmental impacts, as systems rebound and defend themselves against stress.

In this hypothetical (although relevant) example of deciding where to put sludge, protecting fish has to be balanced against protecting humans. Is the probability of avoiding one excess premature death worth reducing the risk to fish in one cubic mile of ocean? A hundred cubic miles? The North Atlantic? Or, with respect to timing, is avoiding one probable excess premature death now worth as much as avoiding one next year? Or avoiding 500, let us say, 1,000 years from now?

Furthermore, how much is avoiding that excess premature death this year worth? A million dollars worth of other desired expenditures or programs? A billion dollars? The gross national product of the state of New York? What life extension as well as life enhancement would those other allocations yield?

On what basis is a lower level of risk to many to be traded off against a higher level of risk to a few? Is each person of equal concern? If so, is it just the number of health effects that is to be minimized, or is the relationship more complex than this? What about disabling or even merely uncomfortable health effects? How are they to be reckoned when the alternative is the possibility of an excess premature death? What is the rate of exchange between colds and cancer?

For those in public service, the temptation is to say that these are, in principle, unanswerable questions and that they cannot be considered together in one decision. Another response might be that the choices are too cruel to have morally acceptable answers; therefore, they should never be presented in stark terms that require individuals to face them—and face themselves after they have chosen. Rather, a veil should be cast over such choices so that the public is not exposed to them and made both uneasy and a knowing party to an essentially immoral decision.

It must be remembered, however, that the sludge has to go somewhere. When it gets there, the fabric of consequences are real, and the trade-offs will have been made. Human lives may have been exchanged for fish; current lives may have been traded off for lives in the future; one array of goods and services and risks will have been experienced while others will not; health risks of one sort will have been distributed in a particular way to a particular population. A set of values will have been summarized in an explicit decision—and somebody made that decision for the rest of the nation's citizens. In making it, that person or persons had to choose among options that exhibited different kinds of goods and bads. In the process, a common basis of comparison was used—whether it was conscious or unconscious, freely admitted or kept secret. Apples and oranges cannot be added, but how many of one must be given up to get how many more of the other can—and I believe should—be reckoned consciously, before a decision is made.

I noted earlier the resistance to a decision process that openly confronts such trade-offs, a process that, however gingerly, puts a "price" on health effects or ecological damage. Again, I can sympathize with that kind of resistance. The rhetoric that supports it strikes a primitive chord and appeals to our childlike longing for a world in which every problem has a solution and that solution is an unalloyed good. Nevertheless, cruel choices have got to be made, and it seems to me surely irresponsible in a public official to make those choices without tracing their consequences to the fullest extent possible. Formal analysis must be brought in for this task. The values to be placed on different ecological, health, economic, social, and personal outcomes at different times are supplied by the decision maker, who is responsible to the political process. The decision can be made on the basis of the balance of the apparent advantage of one option over another. This response is my incomplete and still unsatisfactory answer to the question posed at the beginning of this paper regarding the nature of a "principled basis" for making environmental decisions.

In terms of the second question posed earlier, I believe legitimacy flows from an acceptance of the decision, or at least of the decision process, by those affected. If this belief is valid, it can be achieved only when the bases of decisions are made explicit and open so that citizens also experience the reality of cruel choices, a policy that offers the possibility of true accountability, should citizens choose to exercise their oversight potential. The opportunity to confer or remove authority is essential to a free and democratic society, and providing the information on which such action may be based is essential in sustaining legitimacy. Moreover, in my judgment, being explicit and open about controversial choices is an exercise in leadership. It leads to a successful, lasting resolution of the case in point and also develops among citizens greater sophistication and understanding to make other decisions in the future.

3
The Politics of Benefit-Cost Analysis

R. SHEP MELNICK

It is hardly surprising that the use of cost-benefit analysis, quantitative risk assessment, and similar analytic tools generates substantial political controversy in the United States. The risks, costs, and benefits under scrutiny are usually difficult to estimate with precision. As one Environmental Protection Agency (EPA) scientist so colorfully put it, "One of the nice things about the environmental standard setting business is that you are always setting the standard at a level where the data is lousy" (quoted in Melnick, 1983:244). Moreover, quantitative analysis frequently spotlights politically and ethically troublesome distributional issues, issues that pit citizen against citizen, nation against nation, and even generation against generation. Sometimes the choices involved are "tragic" in that they require us to decide not just who shall live but who must die (Calabresi and Bobbitt, 1978). Such analysis, in short, is never a purely technical undertaking; it exposes rather than resolves hard political choices. For this reason, most practitioners insist that benefit-cost analysis is a "decision helping" rather than a "decision making" tool.

Adding fuel to these regulatory controversies are several beliefs that are particularly strong and widely shared in the United States. This country has a strong streak of populism that equates big—whether it be business or government—with bad. Indeed, trust in both business and government decreased precipitously in the 1970s. American political culture also puts a premium on procedures that offer a wide variety of interest groups and citizens the opportunity to participate in decision making. The United States' peculiar governmental institutions—particularly its independent and energetic legislative and judicial branches—reflect and reinforce these beliefs. It is not surprising, therefore, that a number of recent studies have shown that

R. Shep Melnick is associate professor of politics at Brandeis University and a member of the associated staff of the Brookings Institution.

environmental regulation is much more adversarial and contentious in the United States than in other advanced industrial democracies (Badaracco, 1985; Brickman, Jasanoff, and Ilgen, 1985; Coppock, 1985; Vogel, 1986).

What is startling, however, is the extent to which many key actors in the U.S. environmental regulation arena claim that decision makers should not even attempt to estimate, compare, or balance risks, costs, and benefits in making regulatory decisions. This absolutist, moralistic rhetoric is a uniquely American phenomenon. Much of our public debate focuses not on such important and difficult questions as how to discount future benefits or how to estimate cancer rates but on whether it is proper (or, as some contend, sinful) to "put a price tag on human life." Consider the following examples.

• In 1981–1982 the Subcommittee on Environmental Protection of the Senate Committee on Environment and Public Works unanimously defeated a proposal to allow EPA to consider cost in setting primary (health-based) air quality standards. It even defeated a proposal to allow EPA to consider cost in setting secondary (welfare-based) air quality standards. One senator explained that, if these proposals had passed, "[w]e would no longer consider acceptable air quality, but a standard we can afford" (*Environment Reporter—Current Developments* [1982] 12:891).

• When EPA decided to consider both cost and extent of risk in setting limits on "hazardous" air pollutants, Congressman Henry Waxman, chairman of the House subcommittee with jurisdiction over the program, castigated the agency for "writing off" endangered individuals and "reduc[ing] human lives to statistics" (*Environment Reporter—Current Developments* [1985]15:306). Environmental groups fought this policy in court, arguing "you can't compromise the decision to set standards with cost considerations. You cannot take into account the cost of control and use that number to argue against controlling pollutants" (*Environment Reporter— Current Developments* [1985]15:865)

• In 1981 the U.S. Supreme Court ruled that the Occupational Safety and Health Act precluded the use of benefit-cost analysis in setting "permissible emission limits" *American Textile Manufacturers* v. *Donovan*, 452 U.S. 490 [1981]). Previously, the D.C. Circuit had found that the "legislative history of the [Clean Air] Act also shows that the Administrator may not consider economic and technological feasibility in setting air quality standards" (*Lead Industries Association* v. *EPA*, 647 F.2d 1130 [1980] at 1149).

The implication of each of these illustrations is that nothing should be allowed to stand in the way of reducing possible health risks—which ultimately means the reduction to zero of human exposure to pollutants linked to disease.

Those who oppose the use of benefit-cost analysis[1] of course, are seldom completely faithful to their "health-only" creed. They talk about economic feasibility rather than cost; they countenance lax enforcement; they create exemptions for special classes of polluters; they encourage— indeed, sometimes even demand—delay, lest the consequences of their general policy become too apparent. For example, in 1983 the very same Congressman Waxman quoted earlier attacked EPA Administrator Anne Gorsuch Burford for imposing sanctions on areas that failed to meet air quality standards for ozone (i.e., smog). He called for a more "flexible" approach and successfully sponsored legislation to extend statutory deadlines once again (Melnick, 1984). This incident highlights the central paradox of the controversy: the widespread hostility to the use of benefit-cost and risk assessment analysis is based on an absolutist health-only position that virtually no one is willing to embrace in the real world. To put it more bluntly, almost no one really believes what many informed people emphatically maintain in public.

Human beings in general and elected officials in particular find it difficult to admit that the policies they support leave some innocent people at risk, especially when that risk is potentially lethal. Yet why is it that health-only legislation and rhetoric are common in the United States but virtually unknown in Western Europe? One would expect just the opposite. The United States is generally seen as more sympathetic to free enterprise, more suspicious of government control, more pragmatic, and more inclined to act slowly and incrementally than are European nations.

The answer, I will argue, lies above all in the structure of U.S. political institutions. Political power in this country is remarkably dispersed. Despite the growth of the national government in recent decades, state and local governments remain important players in the area of environmental regulation. The U.S. Congress is by far the most powerful and active legislative body in the world. While other legislatures have become subservient to the executive, the American Congress underwent a resurgence in the late 1960s and early 1970s. Central to this reassertion of congressional authority was further decentralization of power, commonly known as the rise of subcommittee government. Nowhere is the influence of Congress or the extent of decentralization more evident than in environmental protection. For example, from 1969 to 1979, Edmund Muskie, the chairman of a Senate subcommittee, had at least as much influence on environmental policy

---

[1] Throughout this paper, I use the term *benefit-cost analysis* as a shorthand for a variety of techniques for quantifying and comparing costs, risks, and benefits. Although the differences among benefit-cost analysis, cost-effectiveness analysis, and various forms of risk assessment are substantial, for the purposes of this paper, those differences are of limited importance. The focus here is the nature of political opposition to *any* form of explicit consideration of cost or degree of risk.

as Presidents Richard Nixon and Jimmy Carter, or EPA Administrators William Ruckelshaus and Douglas Costle. In parliamentary systems, few members of the "loyal opposition" retain so much power.

Much the same can be said of the activity of the judiciary. Hardly any major environmental policy escapes close scrutiny by the courts. Federal judges have issued hundreds of decisions shaping regulatory policy. In the words of Brickman, Jasanoff, and Ilgen (1985:46), "[e]ven a casual observer is struck by the vastly lower level of judicial involvement in European regulatory processes." Several environmental groups, most notably the Natural Resources Defense Council and the Environmental Defense Fund, have used their success in litigation to become major participants in national policy making. In the United States, each level and branch of government offers access to a wide variety of groups, including corporations, trade associations, labor unions, professional associations, and intergovernmental lobbies, as well as environmental groups.

Dispersion of power has three important consequences for environmental policy making. First and foremost is the dispersion of responsibility. In the United States, it is easy to shift the blame for nearly everything to someone else (see Weaver, 1987). Second, because no one controls the entire policy-making process, each participant tries to squeeze as much as possible out of the limited portion he or she controls. Third, given the complexity of the entire process, it is difficult to see the connection between the decisions of each participant and eventual outcomes.

These factors in turn affect the receptivity of political actors to benefit-cost analysis. Politicians find it very tempting to take an absolutist, health-only stance when those who will actually impose restrictions on employers, employees, and consumers are located in a different branch of government. The temptation becomes nearly irresistible when the other branch is nominally controlled by the rival political party. Moreover, each participant reacts to the perceived biases of the others. Congress exaggerates its goals because it expects the executive branch to water them down; the executive branch does in fact water them down, in part because it considers Congress's goals to be hopelessly irrational. Given the distrust created by these self-fulfilling prophecies, it is hard to engineer compromise, especially when compromise requires a long sequence of decisions. In addition, the difficulty of connecting particular governmental decisions with real-world outcomes makes it hard to convince anyone that exaggerated, absolutist demands have unfortunate long-term consequences.

These problems by no means prove that the European approach is superior to ours. The United States spends more on environmental protection—measured both in total dollars and as a percentage of its gross national product (GNP)—than any other industrial nation. The key unanswered question is whether the United States gets more "bang" for these environmental "bucks." Most scholars who have addressed this issue have retreated to agnosticism. For example, in his extensive comparative study of American and British regulation, David Vogel (1986:146) notes that "it is difficult to determine the comparative effectiveness of governmental regulations in different countries"; he concludes that "[o]n balance, neither nation's regulatory policies have been significantly more or less effective than the other's: both have had some notable achievements and some conspicuous failures." (See also Brickman, Jasanoff, and Ilgen, 1985:313-314.) One cannot even say with confidence that moving to a less adversarial system would increase receptivity to benefit-cost analysis. Indeed, the very informality of European policy making militates against the use of such formal analysis. The European example stands not so much as a model for emulation as a reminder of the peculiarities of American politics.

ALTERNATIVE EXPLANATIONS

Most of this paper is devoted to explaining the incentives and strategies of congressmen, judges, agency officials, and environmental advocates. Before delving into this institutional and legal analysis, however, it is worth considering three simpler, more commonly heard explanations for the American antipathy to benefit-cost analysis. Each explanation has some merit, but each is also seriously incomplete.

Thinking Like Lawyers

Benefit-cost analysis is a tool devised by economists. Yet lawyers dominate Congress, the courts, and the upper echelons of most regulatory agencies. Economists think in terms of opportunity costs and incentives; lawyers think in terms of rules and penalties and of defeating their adversary (Schultze, 1977; Rhoads, 1985). Not only are lawyers suspicious of techniques they do not understand, but they are unwilling to accept a process they cannot control.

The predominance of lawyers in Congress most likely explains the heavy reliance on "command and control" regulation throughout the 1970s. Still, the thinking-like-a-lawyer argument grows less convincing with every passing year. Congress has embraced the use of benefit-cost analysis for water projects and other programs. Moreover, most of the key concepts behind benefit-cost analysis (e.g., opportunity costs and the impossibility of eliminating all risks) are all too familiar to those who make decisions about the federal budget. In other words, the language of economists is foreign neither to the world of politics nor to the world of the law.

Just as importantly, the number and influence of economists and "policy analysts" is increasing both in Congress and in the executive branch.

As Derthick and Quirk (1985) show, deregulation of airlines, trucking, and telecommunications occurred in the 1970s largely because congressmen, presidents, and regulators accepted the arguments presented by this growing herd of economists. If economists have prevailed elsewhere, why not in the realm of environmental policy? The answer lies in the nature of political incentives.

Ravenous Bureaucrats

Many critics of health and safety regulatory policy blame overzealous bureaucrats for excessive regulation. There is no more common theme among small businessmen, Republicans, or contributors to *Regulation* magazine. Environmental protection agencies, according to these critics, attract people who are single-mindedly committed to protecting the environment. Bureaucrats seek to expand their empire. Government officials revel in red tape. And so on.

Regulatory agencies such as EPA and the Occupational Safety and Health Administration (OSHA) undoubtedly attract personnel who accept the mission of their organizations. It is not hard to find officials at a variety of levels who can be described as zealots. At the same time, however, some regulatory agencies have spawned efforts to expand the use of benefit-cost analysis and to find other ways to balance environmental protection, economic growth, and energy production. In several instances, EPA has turned to benefit-cost analysis despite criticism from Congress.

Most regulatory agencies are internally diverse, numbering economists and political executives, as well as lawyers, engineers, and scientists, among their staffs. Political executives must take responsibility for the consequences of agency decisions—economic as well as environmental. Agency economists spend a good deal of their time estimating the economic consequences of regulatory decisions and responding to arguments put forth by economists outside the agency. These two factors—political responsibility and the professional norms of economists—sometimes lead agencies to embrace forms of analysis that are heartily disliked by their allies in Congress and by environmental groups.

Media Hype

A number of recent studies have shown that the alleged anti-business bias of the press is more than just a figment of Jesse Helms's fertile imagination. Public perceptions of environmental risks are to a large extent shaped by the media, which in turn tend to dramatize and exaggerate those health risks that can be personalized and photographed. Publicizing a new risk or emphasizing an existing one creates intense political pressure to act.

In a "crisis," few politicians dare ask what the effort will cost. The press then turns to other matters, but regulations remain in place—one part of the "regulatory ratchet" described by Bardach and Kagan (1982:Chap. 7).

In his description of press coverage of a haphazardly researched report on the Love Canal, Marc Landy illustrated how some environmental issues get placed at the top of the national agenda:

> The report's sensational language, coupled with the excellent photo opportunities presented by mauve lawns, chartreuse basement walls, and irate residents, aroused the news media from its late summer torpor. Love Canal became the leading national news story for days on end. In the following months, documentaries appeared on the networks, *Time* did a cover story, and Jane Fonda paid a tearful visit. (1986:60)

Landy pointed out an important asymmetry in press coverage: reports on health dangers receive considerable attention; more careful examinations of the reports' reliability do not. Another study of the media and regulation (Rothman and Lichter, 1987) has shown that, although the general public views nuclear energy as quite dangerous, most scientists familiar with nuclear power (including those with no financial ties to the industry) consider it relatively safe. The media, Rothman and Lichter argue, have fostered this public perception by mistakenly implying that the scientific community is closely divided on the issue, by devoting disproportionate attention to the most extreme opponents of nuclear power, and by giving more credibility to scientists affiliated with environmental groups than to those affiliated with business or government.

In addition to highlighting health risks, the media are eager to discover scandal. In environmental regulation, scandal usually means exposing "undue" industry influence—political pressure that results in inadequate protection of public health. Some reporters interpret any overt consideration of cost as evidence of undue industry influence. For example, Martin Tolchin, who covers Congress and regulatory affairs for the *New York Times*, makes no effort to hide his contempt for "the spurious standards of cost-benefit analysis, a theory whose flaws unfold as soon as they are held up to public scrutiny" (Tolchin and Tolchin, 1983:124). "This kind of decision making," he and his coauthor assert, "has no place in the public sector" (1983:141). The policies and careers of James Watt, Rita Lavelle, and Anne Gorsuch Burford did not create these deep suspicions of regulatory "capture," but they did much to confirm them.

Reporters, nonetheless, are equal opportunity scandalmongers. In the mid-1970s, newspapers were full of OSHA "horror stories." Local papers are particularly quick to jump on EPA for being too rigid and single-minded in applying emission rules to local industries and municipalities. What seems to characterize the press above all is an eagerness to find fault with whoever appears to be powerful. In the words of Walter Cronkite, "As far as the leftist thing is concerned, that I think is something that comes from the nature of a journalist's work. . . . I think they're inclined to side with humanity rather than with authority and institutions" (quoted in Rothman, 1979:364). Thus, regulators have good reason to believe that, in the eyes of the press, they are damned if they do and damned if they don't. Like nearly everyone else, the press places conflicting demands on the regulatory system.

Public Opinion and Political Culture

A recent EPA study found that the agency's "overall priorities appear more closely aligned with public opinion than with estimated risks" (*Environment Reporter—Current Developments* [1987]17:1823). Public opinion polls continue to show remarkably strong support for environmental programs, particularly those that seek to protect public health (Mitchell, 1984). A 1986 poll, for example, found that 66 percent of its sample agreed with the statement that "protecting the environment is so important that requirements and standards cannot be too high, and continued improvements must be made *regardless of cost*" (Lipset, 1986). A 1981 Harris poll found that 80 percent of the public opposed any relaxation of the Clean Air Act; 65 percent opposed any cost-based constraints on health standards (Melnick, 1983:38). With evidence such as this, it is not surprising that several participants in this conference have argued that the public "demands" strict regulation of environmental hazards.

Why has public support for environmental protection remained so strong? One reason is that environmental programs offer benefits to a wide variety of groups: upper-middle-class hikers, workers in hazardous industries, members of minority groups concentrated in polluted urban centers, suburbanites hoping to protect property values, and business firms who benefit in one way or another from pollution control. Equally important is the fact that the average citizen seldom directly experiences the cost of environmental regulation. It is comforting to believe that somehow corporations (such as General Motors) or wealthy families (such as the DuPonts) rather than consumers and employees will pay for environmental protection. When costs are imposed on private individuals—as they were with transportation controls in the mid-1970s, the proposed ban on saccharine, and interlocking seatbelts—the public response is usually overwhelmingly negative.

Since the public generally believes that business pays for environmental protection, it is not surprising that support for environmental regulation is inversely proportional to confidence in business. Since the mid-1960s, such confidence, particularly in big business, has plummeted. According to

Lipset and Schneider (1986), in 1966, 55 percent of the public expressed "a great deal of confidence in the people running our major companies"; in 1984, only 19 percent shared this view. In 1985, 73 percent of the public believed that "there is too much power concentrated in the hands of a few large companies for the good of the nation." The United States may have no socialist tradition, but it has a populist tradition that expresses many of the same concerns. In his comparative study, David Vogel found that

> the debate over environmental regulation represents a contemporary version of American populism: the interests of "big business" in production and pollution were contrasted with those of the "people" in the preservation of the ecosystem. . . .Threats to the public's health and safety have not been seen, as they are in Britain, as an inevitable component of production and consumption in a highly industrialized and affluent society; rather they have become identified with the profit motive of America's largest firms. (1986:254)

The brief "revolt against regulation" experienced during 1978–1982 appears to have resulted more from decreasing trust in government than from increasing trust in business. Ironically, Ronald Reagan has helped to build trust in government while Ivan Boesky, Michael Millken, and their compatriots have further eroded trust in business.

The public's perception of environmental issues—above all, its perceptions of the nature and distribution of costs and risks—largely determines its response to polling questions that are often misleadingly simplistic. Although environmentalists often argue that regulatory policy should simply respond to public demand (jettisoning analysis in the process), proponents of benefit-cost analysis maintain that the public must be educated about the true nature of the choices faced by policy makers. As Milton Russell (in this volume) puts it, "to obfuscate inevitable choices is to violate the premise of a government based on the consent of the governed. . . ."

Members of Congress, judges, presidents, the public, and the press all make multiple demands of government. Americans want to avoid war but stand up to the nation's enemies; to fund a variety of programs but reduce taxes and the deficit; to encourage broad participation but avoid regulatory delay; to promote economic growth but refrain from harming the ecosystem. That government officials would like to avoid the hard choices necessitated by these conflicting demands is undeniable. Yet in most cases, certain features of the policy-making system—above all, the need to pass a budget—force them to choose. In contrast, regulatory politics has no such unifying, choice-forcing mechanism readily available. Some writers have advocated a "regulatory budget" to bring greater coherence and responsibility to the regulatory process (Litan and Nordhaus, 1983:Chap. 6). One major problem such proposals face is that many participants gain significant benefits from the existing arrangements. The remainder of this paper focuses on the institutional incentives and strategies of those most opposed to the use of benefit-cost analysis.

CONGRESS: KEYSTONE OF THE ENVIRONMENTAL ESTABLISHMENT

The starting point of any discussion of the politics of environmental protection must be the fact that all of the major environmental statutes of the 1970s were the product of congressional rather than presidential initiation. This simple fact has a number of important consequences.

First, members of Congress, especially subcommittee chairmen, consider the Clean Air Act, the Clean Water Act, the Superfund Act, and the other environmental legislation *their* laws. They advocated action when the president was lukewarm or even hostile. To such key figures as Edmund Muskie, Robert Stafford, Paul Rogers, Henry Waxman, and James Florio, statutory intent means *their* intent. They have devoted much of their time to these issues because they consider environmental protection to be a particularly noble and popular cause. In this respect, they are typical of nearly all those who have chosen to sit on environmental protection subcommittees. Here—as in many other policy areas—self-selection creates a bias that administrators ignore at their peril (see, for example, Shepsle, 1978:Chap. 10). John Mendeloff has found that, of the dozens of oversight hearings on health and safety regulation, "all but four featured criticisms that agencies had been too lax" (1987:7–59).

Second, the difficulty of creating broad new programs without presidential leadership requires members of Congress to uncover "crises" that command media attention and demand immediate action. Presidents can attract attention for a handful of legislative initiatives simply because they are president. Members of Congress command media attention only when they uncover scandals or dramatic, life-threatening problems requiring tough, comprehensive solutions. Thus, to be bought, environmental programs must often be oversold.

Third, congressional initiatives on environmental protection are part and parcel of the broad reassertion of congressional power that began in the late 1960s. Congress has claimed that it, rather than the "imperial presidency," should set national policy. Moreover, Congress has declared that federal bureaucracy is too slow, too parochial, and too receptive to the influence of business to deal effectively with environmental problems. The detailed, "action-forcing" statutes passed in the 1970s were founded on a deep distrust of the executive branch and on the conceit that statutory language could provide definitive answers to almost all policy questions (see, for example, Florio, 1986).

On the latter score, congressional entrepreneurs were clearly mistaken. EPA and other regulatory agencies were given little usable guidance on how to set air quality standards, new source performance standards, effluent guidelines, and the like. Given the amount of money at stake, it is not surprising that presidents have sought to have some part in agency decision making. President Nixon initiated the "Quality of Life Review," and President Ford added "Inflation Impact Statements." President Carter created the Regulatory Analysis Review Group, which provided detailed analyses of major environmental rules. The Reagan administration's efforts to increase substantially the power of the Office of Management and Budget (OMB) and to insist on the use of benefit-cost analysis except when expressly prohibited by law are the latest and most extensive attempts by the White House to influence regulatory policy. All of these regulatory review measures received harsh criticism from Capitol Hill.

OMB: The Eye of the Storm

Controversy over the use of benefit-cost analysis is thus intertwined with more than 20 years of legislative-executive conflict. Members of Congress understandably associate benefit-cost analysis with hostile OMB economists seeking to relax environmental standards. They see such analytic techniques as little more than Trojan horses carrying industry lobbyists. Conversely, White House and OMB officials view congressional hostility to benefit-cost analysis as further evidence of congressional demagoguery and stubbornness, and an unwillingness to admit that it is impossible to create a risk-free world. Those conducting "regulatory reviews" believe that benefit-cost analysis partially corrects the unbalanced policies advocated by influential members of Congress; members of Congress, on the other hand, see it as a form of regulatory impoundment.

The contrasting perspectives of Congress and the White House spring from two sources. The first is partisanship: for 17 of the past 21 years, the Democratic Party has dominated Congress and the Republican Party has controlled the presidency. (Some Democrats would add that, in the second half of his term, President Carter acted more like a Republican than a Democrat.) Republicans tend to be more suspicious of government control than are Democrats; Democrats see environmental protection as a good issue to use against Republican presidents. The second cause is institutional. As noted earlier, the most vocal members of Congress are those who are most thoroughly devoted to environmental protection. Moreover, given the broad appeal but low salience of environmental issues for voters, most members of Congress discover that "a pro-environmental voting record can only help, not hurt, at reelection time" (Mitchell, 1984:68). Playing it safe—the strategy of most incumbents—means not appearing to favor dirty air, dirty water, or hazardous waste dumps. Presidential popularity, in contrast, depends primarily on the performance of the economy. To the extent that regulation retards economic growth, it is a threat to a president's success.

Whenever OMB uses benefit-cost analysis to justify environmental standards that are weaker than those favored by agency personnel, congressional outrage is inevitable and, on the surface at least, readily understandable. Yet to what, precisely, does Congress object? Different members tend to give different answers. Seldom is the primary issue the particular set of assumptions used by OMB. Let us take, for example, the lengthy report on asbestos control issued by the House Energy and Commerce Committee's Subcommittee on Oversight and Investigations (U.S. Congress, House, 1985). Only 10 pages of this caustic, 140-page report deal with the content of OMB's analysis, and even there the report suggests no alternatives to OMB's assumptions.

The most frequently voiced criticism of the regulatory review practices of the Reagan administration is that OMB hides behind closed doors. In the words of John Dingell, one of OMB's most persistent critics, "Congressional directives, including those designed to protect the public health and the environment, can be easily circumvented in a review process which is shrouded in secrecy, unbounded by statutory constraints, and accountable to no one." OMB's "secret and heavy handed interference" undermines Congress's "carefully crafted procedures," which were designed "to insure that all interested parties. . . participate on an equal footing" (U.S. Congress, House, 1985:iv). OMB frequently has been accused of serving as a "conduit" for private communications from industry lobbyists. However, the recent agreement between OIRA and key members of Congress on the disclosure of OMB activity could take the steam out of this part of the debate (*Congressional Quarterly Weekly Report*, June 21, 1986:1409).

Some members of Congress go one step further. They say they have little trust in OMB or in other non-agency personnel to produce fair benefit-cost analyses, however open the process. Just as OMB's budgetary mission is to cut spending, its regulatory mission is slowing or even reversing the growth of regulation. As former OIRA Deputy Director James Tozzi explained, OMB reviewers "start with the idea, 'Do you really need this reg?' People say, 'That's such a negative view,' but I say that's a good role for them" (*National Journal*, May 30, 1987:1406). The fact that OMB's principal mechanism for controlling agency action is its ability to delay or veto administrative rules compounds the problem. Benefit-cost analysis becomes a rationale for stalling and weakening regulations, never for initiating or strengthening them. Given the amount of discretion involved in conducting benefit-cost analysis, this institutional-bias argument against OMB review has significant force. Even those individuals within regulatory

agencies who most fervently advocate the use of benefit-cost analysis have
come to resent OMB review. For this reason, it is important to distinguish
between analysis per se and its use by OMB.

The Health-Only Canard

Still other congressional leaders argue that any balancing of costs
and benefits—whether done by OMB or agency officials—conflicts with
policies that have already been established by Congress. A number of
statutes fail to include cost as factor that agencies should take into account;
others mention feasibility rather than cost; and a few specifically limit the
agency's deliberations to health concerns. The sponsors of these legislative
provisions usually maintain that the agency is to pursue a strict policy of
protecting the public health. Taking these statements at face value, courts
have forbidden agencies to use benefit-cost analysis under some statutes
(see the later section entitled "The Federal Courts").

What is to be made of health-only commands? It should be noted
that the most emphatic health-only statements appear not in the statutes
themselves but in committee reports, floor statements offered by sponsors,
and subsequent oversight hearings. No votes are taken on these indices
of "congressional intent," and, at least according to traditional canons of
statutory interpretation, they are not legally binding.

Indeed, most statutory language on standard setting is remarkably
vague. For example, the Clean Air Act (P.L. 91-609, 1970), which many
people claim embodies the health-only approach, directs EPA to set primary
air quality standards that "protect the public health" with "an adequate
margin of safety" and to establish hazardous emission limits that "provide
an ample margin of safety to protect the public health." What does "safe"
mean? It could mean (as Senate Report 91-1196[1970] suggests) setting
standards at health-effect "thresholds"; that is, at levels below which there
are no observable adverse consequences even for sensitive individuals.
Almost everyone agrees, however, that there are no thresholds—other than
zero—for most pollutants. Does this mean that EPA should set standards at
zero or at natural background levels? It is hard to find elected officials who
take this position. But otherwise the health-only interpretation frequently
given to this and similar statutes becomes meaningless—or, to use the
surprisingly frank language of a House report, a "myth."

When it comes to balancing health benefits against cost, many congressmen exhibit a deep-seated schizophrenia (or hypocrisy, depending on
how charitable one wants to be). This attitude was graphically illustrated in
the remarks of Senate staff member Curtis Moore, who presented the congressional perspective at the conference. On the one hand, Moore argued
that emitting potentially harmful pollutants is as morally wrong as slavery

(conference transcript:85) and murder (conference transcript:92 and 418). Consequently, such action should be prohibited regardless of cost:

> The fact of the matter is, whether you like it or not, the American people don't think a person has a right to take their health or their lives because it saves them money. . .and I will tell you flat out that this [benefit-cost analysis] discipline has been used for one reason, and one reason only, and that is to avoid regulation and to save the industry costs. (conference transcript:492)

On the other hand, Moore insisted that "Congress does, in fact, use cost-benefit analysis and risk-benefit analysis in making its decisions" (conference transcript:418). Indeed, "that is the place to do it, it is where you are legislating" (conference transcript:90). After all, "if you happen to be in West Virginia where you can lose your job along with 60,000 other miners where EPA changes the base case analysis, that job is pretty important to you" (conference transcript:83).

If Congress did in fact wish to launch a moral crusade against all health-endangering pollution, it could easily do so. It could ban specific pollutants. It could tell EPA and OSHA to set standards at zero or natural background levels. It could also increase appropriations for enforcement and refuse to extend compliance schedules for polluters. It is important to remember that, despite the congressional criticism heaped on the Reagan administration, it was Congress that approved substantial budget reductions for EPA. As numerous studies indicate, Congress has a variety of techniques for controlling agency discretion (Weingast and Moran, 1982; Moe, 1985; Aberbach, 1987; McCubbins, Noll, and Weingast, 1987). That it has not taken more aggressive action reflects its ambivalence rather than its powerlessness.

In practice, those who support a health-only approach advocate standards that are (a) somewhat more stringent than anyone would consider reasonable to meet in the next few years but (b) not so severe as to promote political backlash. Setting standards that industry can meet today is not "technology-forcing"; such standards do not put polluters and regulators on the defensive. Instituting weaker standards or even strict standards without ambitious deadlines, to quote the dissenting opinion of three members of the National Commission on Air Quality, "would legitimize the perpetual failure to provide healthful air quality" (1981:5–36). Yet standards that put people out of work, dramatically increase consumer costs, or visibly restrict individual freedom threaten to destroy the politically crucial myth that business rather than citizens pays for environmental protection.

This strategy is particularly attractive to those who do not themselves set standards or impose sanctions but who can garner political benefit by criticizing those who do. The executive branch is put in the unhappy position of issuing orders to polluters and of condoning some degree of non-compliance. Congress can take credit for passing bold, technology-forcing legislation, for uncovering administrative ineffectiveness and cowardice, and even for inducing administrators to be more "reasonable" with local employers. In the words of Brickman, Jasanoff, and Ilgen, Congress has not only "laid out impossibly optimistic goals" but established deadlines "that fall somewhere between the merely unrealistic and the wholly fantastic. Yet if goals are not met in timely fashion, Congress bears no direct responsibility. Indeed. . .legislative oversight provides Congress an unparalleled means of making political capital out of agency failure" (1985:72; see also Fiorina, 1977; Melnick, 1983:Chap. 10) Although Congress may have legitimate questions about the current use of benefit-cost analysis, especially by OMB, much of its opposition is an elaborate form of political posturing.

THE FEDERAL COURTS

For two decades the federal courts have been deeply involved in nearly every aspect of environmental policy making. Environmental issues arrived on the national agenda just as administrative law was undergoing a fundamental reformation (Stewart, 1975; Melnick, 1983). Federal judges have not only insisted that agencies follow elaborate new rule-making procedures but have determined the legislative "intent" behind dozens of statutory provisions and ordered administrators to undertake scores of "nondiscretionary duties." Volume upon volume of legal commentary has been devoted to describing these developments. Two central questions are important for discussion here. First, how have judicially mandated procedures affected policy making? Second, have the courts read particular statutes to require, allow, or preclude the use of benefit-cost analysis? I will argue that in the past the federal courts—particularly the D.C. Circuit—strengthened the position of the congressional committee members described earlier, but that this pattern may be changing.

Rule-Making Procedures

Since Judge Bazelon of the D.C. Circuit first announced the arrival of "a new era in the long and fruitful collaboration of administrative agencies and reviewing courts" in 1971 (*EDF* v. *Ruckelshaus*, 439 §2d584 D.C.C. 1971), the federal courts have reshaped the rule-making process to promote both technical accuracy and representational fairness. To achieve these goals the courts have read a number of new requirements into the notice-and-comment rule-making provisions of the Administrative Procedures Act (P.L. 79-404, 1946).

When agencies propose new regulations, they must make public the data, methodologies, and arguments on which their proposals rely. Not only must they invite comment on this material, but they must also respond to all "significant" criticism. Subsequently, they must provide a detailed explanation of how they arrived at their final rule. All of this information must be compiled in a record that can be reviewed by the appropriate court. The reviewing court will take a "hard look" at the record, insisting that the agency "articulate with reasonable clarity its reason for decision and identify the significance of the crucial facts, a course that tends to assure that the agency's policies effectuate general standards, applied without unreasonable discrimination" (*Greater Boston Television Corp.* v. *Federal Communications Commission*, 444 F.2d 841 [D.C. Cir., 1970] at 851). Courts have insisted upon undertaking such "searching and careful" reviews even when such review requires immersion in highly technical material.

The objective of ensuring participation by all affected interests goes hand-in-hand with the goal of ensuring adequate technical analysis. In part the former serves the latter: when all can speak, more information and alternatives are presented for consideration. Yet the concerted efforts of the courts to open the door to participation by such nontraditional participants as environmentalists, civil rights organizations, and consumer groups also reflected judicial concern about the political biases of administrative agencies. In the late 1960s the courts began to complain that administrators were focusing too narrowly on the accepted missions of their agencies and losing sight of the public interest in the process. At best, according to the courts, agencies were unimaginative; at worst, they had been captured by powerful industrial interests. Led by the D.C. Circuit, the courts sought to open up these "iron triangles" by giving a variety of interests the chance to be heard. The "reformation" of administrative law, as Richard Stewart has explained (1975:1712), "changed the focus of judicial review. . .so that its dominant purpose is no longer the prevention of unauthorized intrusion on private autonomy, but the assurance of fair representation for all affected interests in the exercise of the legislative power delegated to agencies." Just as the technical components of agency decision making would be reviewed by the courts for accuracy, the more political components would be scrutinized for fairness and breadth of view.

These judicial developments culminated in the demand for what Martin Shapiro has called "synoptic decisionmaking," a process that "requires all facts to be known, all alternatives to be considered, all values to be identified and placed in an order of priorities and that then selects the alternative that best achieves the values given the facts" (1986:466; see also Diver, 1981). On the surface, at least, this process would seem to encourage more thorough analysis of scientific evidence and more careful weighing of the costs and benefits of regulating. Some of the standards struck down in court were, indeed, seriously flawed. (One example is the air quality standard that was invalidated in *Kennecott Copper Corp.* v. *EPA*, 462 F.2d 846 [D.C. Cir., 1972]). Today, agency anticipation of judicial review makes such shoddy use of evidence unlikely. For this reason the rule-making procedures devised by the courts have been widely praised. Unfortunately, there have been only a few empirical studies of the substantive consequences of these procedural mandates.

There can be no doubt that the amount of evidence and analysis that goes into writing environmental regulations is far greater today than it was in 1970. Judicial requirements have contributed to this increase, but they were only one of several factors at work. The environmental legislation passed in the early 1970s gave the fledgling EPA only a few months to make scores of major decisions. Few states, businesses, or environmental groups realized the importance of these regulations or bothered to submit comments to the agency. Now, when regulations are being promulgated, all sides recognize what is at stake and devote enormous effort to marshaling evidence to support their positions. Not only have EPA and other regulatory agencies built larger and more sophisticated staffs, but they have come to realize the political importance of offering elaborate justifications for their rules.

Of course, enlargement of the rule-making record is not an unmixed blessing. As records have grown more and more massive, agencies have become more wary of judicial reversal. A vicious cycle soon develops: the more effort an agency puts into building a record, the more it fears seeing that effort going to waste, and the more effort it makes to cover all possible bases. This activity not only adds to regulatory delay but makes it less likely that regulators will adjust their standards in the light of new evidence. In this way, judicial review adds to the rigidity for which American regulation is famous.

Perhaps more significant is the subtle way in which court rulings have influenced the types of experts and evidence upon which agencies rely. Serge Taylor has shown that court decisions under the National Environmental Policy Act (P.L. 91-190, 1970) helped bring into the Corps of Engineers and the Forest Service new breeds of specialists who made these development-oriented agencies more sensitive to the environmental consequences of their actions (1984:Chap. 12) Within EPA, court decisions have tended to increase the influence of those responsible for interpreting evidence on health effects—not to mention, of course, the lawyers who construct the agency's arguments for presentation to reviewing courts. To state the case more bluntly, the judiciary has given added authority to the naive belief that the "proper" standard will eventually emerge if one collects enough scientific data and stares at it long and conscientiously (Melnick, 1983:Chap. 8; Coppock, 1985).

Conversely, if court procedures have produced a category of bureaucratic losers, it is political executives. "The demand for synoptic deliberation," Shapiro points out, "encourages agencies to disguise exercises of discretion as exercises of objective synopticism" (1986). To overturn the recommendations of civil servants or to respond to the political agenda of a new administration is to emit what the D.C. Circuit called "danger signals," justifying heightened judicial scrutiny (*State Farm Mutual* v. *Department of Transportation*, 680 F.2d 206 [D.C. Cir., 1982] at 228–230; Garland, 1985:517–521). Ironically, although in the 1960s and early 1970s the courts stressed the need to strengthen bureaucratic accountability, more recently, they have emphasized the danger of "political" interference with the administrative process.

Relying on this logic, environmental litigants have sought to place tight legal constraints on OMB review of agency rules. OMB, they have argued, should do nothing more than submit comments that become part of the formal record. The courts, however, have not adopted this position. In *Sierra Club* v. *Costle* (657 F.2d 298 [1981]), the D.C. Circuit stated, "[w]e do not believe that Congress intended that the courts convert informal rulemaking into a rarified technocratic process, unaffected by political considerations or the presence of Presidential power" (at 408). Judge Wald, a former "public interest" lawyer, wrote the opinion in the case and presents strong constitutional and practical arguments for extensive White House review of agency rules:

> The authority of the President to control and supervise executive policy-making is derived from the Constitution; the desirability of such control is demonstrable from the practical realities of administrative rulemaking. Regulations such as those involved here demand a careful weighing of cost, environmental, and energy considerations. They also have broad implications for national economic policy. Our form of government simply could not function effectively or rationally if key executive policymakers were isolated from each other and from the Chief Executive. Single mission agencies do not always have the answers to complex regulatory problems. An over-worked administrator exposed on a 24-hour basis to a dedicated but zealous staff needs to know the arguments and ideas of policymakers in other agencies as well as in the White House. (at 406, notes omitted)

The courts have held, in effect, that regulatory review will be judged by its fruits: to the extent that OMB pressure leads to a rapid reversal of agency policy or produces open conflict between the agency and OMB, reviewing courts will take a particularly "hard look" at the adequacy of the agency's formal justification for its rule. This "hard look," however, will not always prove fatal for the regulation in question.

The institutional and policy consequences of the demand for "synoptic" decision making (Shapiro, 1986) are difficult to predict with precision for the simple reason that synoptic decision making is impossible to perform in the real world. Judges must decide whether administrators have done a "good enough" job. Not surprisingly, different judges may come to wildly different conclusions. Circuit courts reviewing the decisions of the National Highway Traffic Safety Administration, for example, have adopted such a demanding evidentiary standard that the agency has virtually abandoned regulation through rule making (Mashaw and Harfst, 1987). The D.C. Circuit, in contrast, has adopted a permissive standard of review for air quality standards under the Clean Air Act. Not only do judges differ in their assessments of the competence and missions of agencies, but the "weight" that judges allow administrators to give to each "factor" depends on the idiosyncratic wording of the statute in question. This divergence among judges leads to the issue of whether the courts have interpreted environmental protection statutes to allow, prohibit, or require the use of benefit-cost analysis.

Reading Statutes

The courts' reading of health and safety statues has been something less than a model of clarity. Obscurity and evasion are evident both in the Supreme Court's two major decisions in the area and in the D.C. Circuit's most recent pronouncements on the regulation of carcinogens. Richard Pierce's general description of judicial review by the current D.C. Circuit applies with a vengeance in this policy area: "Assessing the likelihood of success in making policy through rulemaking increasingly resembles the process of predicting the result of a lottery." (1988:302)

In *Industrial Union Department, AFL-CIO* v. *American Petroleum Institute* (448 U.S. 6-7 [1980]), a fragmented Supreme Court struck down OSHA's benzene standard, claiming that OSHA had not shown "that the toxic substance in question poses a significant health risk." The majority stressed that the Occupational Safety and Health Act (P.L. 91-596, 1970) does not demand a "risk-free" work environment: "In the absence of a clear mandate in the Act, it is unreasonable to assume that Congress intended to give the Secretary the unprecedented power over American industry that would result" (p. 645). Yet the Court refused to "express any opinion on the more difficult question of what factual determination would warrant a conclusion that significant risks are present" (p. 659). More to the point, the Court had "no occasion to determine whether costs must be weighed against benefits" (p. 640). Only Justice Powell, writing a concurring opinion, found that the act "requires the agency to determine that the economic effects of its standard bear a reasonable relationship to the expected benefits" (p. 670).

Less than a year later, the four dissenters in the benzene case joined with Justice Stevens, the author of the benzene opinion, to rule that the "feasibility" requirement of the Occupational Safety and Health Act precludes use of benefit-cost analysis. The crux of the Court's argument in *American Textile Manufacturers Institute* v. *Donovan* (452 U.S. 490 [1981]) was the following:

> Congress itself defined the basic relationship between cost and benefits, by placing the "benefit" of worker health above all other considerations save those making attainment of this "benefit" achievable. Any standard based on a balancing of costs and benefits by the Secretary that strikes a different balance than struck by Congress would be inconsistent with the command set forth [in the act]. Thus, cost-benefit analysis by OSHA is not required by the statute because feasibility analysis is. . . . When Congress has intended that an agency engage in cost-benefit analysis, it has clearly indicated such intent on the face of the statute (pp. 509-510).

As this excerpt indicates, out went "significance" and in came "feasibility" as the touchstone of standard setting. One suspects that both terms are broad enough to allow administrators to justify nearly any decision they reach, provided they know which rule the courts will apply.

The Supreme Court's lack of attention to most regulatory statutes, coupled with its inability to offer clear directives to the lower courts, leaves the D.C. Circuit as the key court for most regulatory agencies. In the words of Justice Scalia (1978:371), "as a practical matter, the D.C. Circuit is something of a resident manager, and the Supreme Court an absentee landlord in administrative law." For many years the D.C. Circuit read regulatory statutes to require highly protective health-only standards. But that position may be changing.

In several important decisions in the late 1970s and early 1980s, the D.C. Circuit created a legal presumption against the use of benefit-cost analysis—or any other consideration of cost—in many forms of standard setting. This presumption was most evident in cases involving carcinogens. In *Environmental Defense Fund* v. *EPA* (598 F.2d 62 [1978]), the court stated:

> An administrator has a "heavy burden" to "explain the basis for his decision to permit the continued use of a chemical known to produce cancer in experimental animals." When firm evidence establishes that a chemical is a carcinogen, statutes generally leave an administrator no alternative but to step in to protect the public health. . . ."Courts have traditionally recognized a special judicial interest in protecting the public health," particularly where "the matter involved is as sensitive and fright-laden as cancer" (p. 88, cites omitted).

The court adopted a similar "precautionary" position in interpreting the Clean Air Act, even if the pollutant is not carcinogenic. In *Lead Industries Association* v. *EPA* (647 F.2d 1130 [1980]), Judge Skelly Wright announced that "the legislative history of the Act shows the Administrator *may not* consider economic or technological feasibility in setting air quality standards; [this] was . . . the result of a deliberate decision by Congress to subordinate such concerns to the achievement of health goals" (p. 1149, emphasis added). The court reaffirmed this position in *American Petroleum Institute* v. *Costle* (665 F.2d 1176 [1981]).

The judges appointed to the D.C. Circuit by President Reagan have been considerably less sympathetic to such arguments than have judges appointed by previous presidents. A 1986 opinion written by Judge Robert Bork allowed EPA to use a very rough form of benefit-cost analysis to set a "hazardous emission" standard for vinyl chloride, despite the facts that vinyl chloride is a carcinogen and the statutory language on "hazardous pollutants" is stronger than the language interpreted by Judge Wright in the Lead Industries decision (*Natural Resources Defense Council* v. *EPA, Environment Reporter—Cases* 25:1106 [D.C. Cir., 1986]). Judge Bork's opinion, in effect, reversed the presumption about consideration of cost:

> The statute brings the Administrator's discretion and judgment to bear on scientific uncertainty. If health were the only permissible consideration, no such discretion would be necessary, for deciding how much uncertainty to allow from a strictly health-based perspective would always lead to the same answer—none. . . . [A]ny decision informed solely by health, but no other, values would require a prohibition of any emissions. Had Congress intended that result, it could very easily have said so by writing a statute that states that no level of emissions shall be allowed as to which there is any uncertainty (p. 1110).

Judge Bork's opinion drew a sharp dissent from Judge Wright, who correctly saw it as a repudiation of previous rulings.[2]

Such doctrinal clarity, though, was too much for the factious D.C. Circuit to bear. An 11-member *en banc* panel issued a revised opinion and

---

[2] Judge Wright's dissenting opinion included an instructive misquotation. He attempted to refute the above-cited argument of Judge Bork by showing that it is possible to set nonzero emission standards without considering cost. To buttress his claim that health effect "thresholds" exist for some pollutants, he stated that "Senator Muskie, reflecting back on the legislative process, has said that the Act is based on the assumption that thresholds of safety exist for some hazardous pollutants" (p. 1127). What Senator Muskie actually said in the hearings cited by Judge Wright is this: "Scientists and doctors have told us that there is no threshold, that any air pollution is harmful. The Clean Air Act is based on the assumption, *although we knew at the time it was inaccurate*, that there is a threshold." What one House subcommittee called the "myth" of thresholds is central to the argument against the consideration of cost in standard setting (Melnick, 1983:Chap. 8). Judge Wright has insisted on clinging to this myth.

sent the vinyl chloride standard back to EPA. This second opinion (also written by Judge Bork) followed the first in rejecting the Natural Resources Defense Council's argument that EPA must set a standard of zero for a nonthreshold pollutant. According to the court, "Since we cannot discern clear congressional intent to preclude consideration of cost and technological feasibility . . . we necessarily find that the Administrator may consider these factors" (*Environment Reporter—Cases* [1987] 26:1263 at 1278). Yet such considerations can come into play only after the Administrator has made "an initial determination of what is 'safe'" (p. 1280). This decision "must be based solely upon the risk to health. The Administrator cannot under any circumstances consider cost and technological feasibility at this stage of the analysis." Cost and feasibility, apparently, can influence only the size of the "margin of safety." But the court also emphasized that "safe" cannot mean "risk-free" and that even during the first stage of analysis the Administrator must use his "expert judgment" to determine what is an "'acceptable' risk to health." The *en banc* opinion, in short, bore all the marks of a report written by a committee. After making a series of contradictory arguments, the court remanded the standard to EPA "for timely reconsideration of the 1977 proposed rule consistent with this opinion" (p. 1281).

Given the intellectual disarray in the Supreme Court and the D.C. Circuit, one should not expect too much consistency from the courts. Although the courts have given EPA and regulatory reviewers more elbow room in such cases as *Sierra Club* v. *Costle* and *Natural Resources Defense Council* v. *EPA*, in a variety of other cases the courts have struck down rules (or recisions of rules) they consider too lenient.[3] In 1985 the D.C. Circuit heard 19 cases involving deregulation. Agencies won 11 and lost 8. In 6 of the 8 the agency lost, the court found the agency's explanation for its policy inadequate (Wald, 1986:537). This trend indicates that administrators who admit to using some form of benefit-cost analysis in setting health and safety standards would be well advised to collect a good deal of support for their position—from legislative histories as well as from more technical data—and to show that they are not merely responding to pressure from OMB, the White House, or industry. All they can do then is hope to face a sympathetic panel on the D.C. Circuit.

---

[3] These cases include the following: *Motor Vehicle Manufacturers Assoc.* v. *State Farm Mutual* (436 U.S. 29 [1983]), *Farmworkers Justice Fund* v. *Brock, Occupational Safety and Health—Cases* (13:1059 [D.C. Cir., 1987]), *Public Citizen Health Research Group* v. *Tyson, Occupational Safety and Health—Cases* (12:1905 [D.C. Cir., 1987]), and the large number of cases listed in Garland (1986) at n.185. Data for 1987 show that the D.C. Circuit approved the ruling of the administrative agency in only 40 percent of the cases it heard (Pierce, 1988:301).

REGULATORY AGENCIES

Two images of regulatory agencies frequently surface in discussions of health and safety regulation. The first stereotype is that of the regulator as zealot. The second is the image of the "captured" agency. The first view encourages the belief that regulators will never voluntarily initiate—or even implement, in good faith—procedures to quantify and compare costs, benefits, and risks. To exaggerate only slightly, this is the view that often pervades OMB. In contrast, the capture version of the story paints benefit-cost analysis as little more than a tool of industry lobbyists and agency use of such analysis as evidence that regulators have "sold out" once again. Like all stereotypes, these two fail to reflect the complexity of political life. It is useful to consider at greater length why administrators are sometimes driven to use benefit-cost analysis and why they remain wary of employing it more fully.

Derthick and Quirk (1985:Chap. 3) have found that in three agencies administering "economic" regulation (the Civil Aeronautics Board, the Federal Communications Commission, and the Interstate Commerce Commission), a number of regulators adopted—even preached—the views of mainstream microeconomics, even though this analysis threatened the very survival of their agencies. Derthick and Quirk found two major causes of this behavior. First, when presidents care to do so, they can usually appoint political executives who share their political views. These appointees can have a significant impact on the agencies they head, even when these views are at odds with the mission of the agency. Second, agencies sometimes house dissidents who become disillusioned with the performance of the agency and seek to change its behavior.

The Reagan administration made unprecedented efforts to ensure that its political appointees were skeptical of or even openly hostile to what it considered social regulation. In a few instances (especially those involving Anne Gorsuch Burford but also Raymond Peck at the National Highway Traffic Safety Administration), the resulting animosity between agency and chief administrator was destructive of both agency morale and regulatory reform. In other cases, less abrasive, more knowledgeable appointees have succeeded in encouraging the greater use of benefit-cost analysis, in part by building up the offices responsible for performing economic analysis. Perhaps the best example of this approach is the Federal Trade Commission under James Miller.

It is important to note, however, that EPA had begun to place greater emphasis on economic analysis well before 1981. During the Carter years the Office of Planning and Management under William Drayton increased both its technical sophistication and its internal political clout. Not only did it benefit from its position as the unit responsible for countering the arguments of Regulatory Analysis Review Group economists, but it began to educate itself and the rest of the agency about the cost and effectiveness of various programs. The hazardous air pollution policy discussed in the preceding section, a policy that departs from EPA's normal health-only stance, took shape during these years. Very few people at EPA wanted to mandate hugely expensive hazardous emission controls to reduce already small risks. Faced with an all-or-nothing choice, most staff viewed nothing as preferable. Because they remained concerned about some of these health risks, however, they sought to broaden the array of choices open to the agency.

There are at least two reasons for believing that regulators, including those in the civil service, will become increasingly sympathetic to the use of techniques for estimating and comparing costs and risks. The first reason is internal. As more and more new jobs are assigned to regulatory agencies and as the complexity of these tasks becomes apparent, regulators will want some indication that they are addressing important problems rather than trivial ones. Ordered by Congress to make everything a priority, regulators must find some non-statutory basis for ranking their tasks. Cost-effectiveness is an obvious candidate.

The second reason involves the mobilization of political support. Two decades of experience with environmental regulation show that it is not easy to change social and economic practices that adversely affect the environment. As costs mount, Congress becomes more ambivalent, and opposition from within the executive branch intensifies. When industry views regulation as ruinous, it pulls out all the stops in its opposition, trying first to block agency rules and then to avoid complying with them. (The history of air pollution regulations for steel mills, smelters, and midwestern utilities clearly illustrates this dreary fact.) To enforce pollution rules, EPA needs the cooperation not just of industry but of state and local governments and federal district court judges as well (Melnick, 1983:Chap. 7). Not only must the agency avoid "going to the well" too often, but it will be severely hampered if it cannot show each of these actors that costs bear some rough resemblance to benefits. Faced with abstract policy questions, the public often advocates paying "any price" for a clean environment; faced with the prospect of actually bearing these costs, most people change their mind.

There remain two important obstacles to the greater use of benefit-cost analysis in regulatory agencies. The first obstacle is the problem of image. In opposing the explicit consideration of cost in setting the ambient standard for airborne lead, two EPA lawyers argued that "[w]e have billed ourselves emphatically of late as a health protection agency. This is an instance where we really need to behave as if we believe our image-making" (quoted in Melnick, 1983:278). To the extent that an agency appears to forsake its role as an advocate for the protection of the environment and of public health, it endangers not only its support on Capitol Hill and among environmental groups but its special place in public opinion. It is better, perhaps, to say "no, never" than to appear to haggle over price.

For some time, EPA has surmounted this obstacle by setting stringent standards and compromising on compliance schedules. For example, in announcing the lead standard mentioned above, EPA conceded that it did "not believe that a major disruption of this industry [smelting] is an acceptable consequence." It promised to "explore every avenue," including the extension of statutory deadlines, "to avoid such an impact while still protecting the public health" (*Environment Reporter—Current Developments* 9:1091). Some compliance schedules are negotiated, violated, and renegotiated. Statutory deadlines are extended time and time again. The worst example is the air quality standard for ozone, attainment of which was first set for 1975, then 1977, then 1985, then 1987—and which will never be met in some cities. This process leads some citizens to ask, "Why is it we still don't have clean air?" and leads others to suspect that EPA does not really mean what it says. Not only does this strategy have perverse economic consequences (especially a strong bias against new forms and sources of pollution), but it breeds cynicism, which makes all regulatory activity more difficult. In short, the image problem reappears in a different form.

The second, more persistent problem is enforcement slippage. Many enforcement officials believe that the only way to get reasonable results is to set more ambitious standards. If you want people to drive 60 miles per hour (mph), then you need to set the speed limit at 50 mph; setting it at 60 mph will lead people to drive 70 mph. (Of course, as Robert Kagan once pointed out to me, if you set it at 10 mph, people will ignore the law altogether.) Enforcement is always a bargaining process, and regulators can only bargain down from a standard. Moreover, industry will often cooperate with state agencies only if it fears that the alternative will be a more onerous federal requirement. Economists who prepare benefit-cost analyses usually make the highly questionable assumption that standards published in the *Federal Register* generate complete compliance. Administrators know better.

In Western Europe and Japan, a less demanding, less adversarial standard-setting process goes hand in hand with more effective enforcement procedures. It seems reasonable to assume that if the United States were to set more lenient standards, it would meet with less resistance in enforcement. Unfortunately, however, reducing demands does not in itself guarantee either more cooperation from industry or more enforcement resources from OMB or Congress. Indeed, the Reagan administration seriously and consciously weakened EPA's enforcement capability at the same time that it sought to loosen standards. This approach makes many people reluctant to endorse any major changes in EPA policy. Those who want to make standards more reasonable must be willing to show that they are committed to aggressive enforcement of the revised requirements.

ENVIRONMENTAL GROUPS

Leaders of environmental groups are clearly among the most vociferous opponents of benefit-cost analysis. Because these groups have, to a remarkable extent, retained their influence despite the "Reagan Revolution," their views count. Yet understanding the nature of their opposition is not always easy. Environmentalists do not distrust all forms of economic analysis: they have supported the use of benefit-cost analysis for water projects and timber sales, as well as the application of marginal cost pricing to electricity. Still, in most other areas, environmentalists have claimed that benefit-cost analysis inevitably underestimates the environmental benefits and overstates the economic costs of a policy. One must ask why, if these mistakes are so clear, environmentalists do not seek to correct them rather than to reject all efforts to compare costs and risks.

There are several plausible explanations for their behavior. First and most obviously, environmental leaders need to maintain the viability of their organizations. Voluntary organizations, especially those depending on contributions raised through direct mail solicitation, must make simple moral appeals to their constituents. "Polluters are killing people and we must make them stop" has much more pizzazz than "let's raise the cost-per-life-saved from $1 million to $2 million." No one wants to abandon the high moral ground. Moreover, as noted earlier, unmet standards provide opportunities for further crusades and lawsuits.

Second, the environmental groups that are most active at the national level—especially the Natural Resources Defense Council, the Environmental Defense Fund, and the Sierra Club—have more influence during the legislation-writing and standard-setting phases of regulation and less influence in enforcement oversight. In addition, they view their information-gathering resources as vastly inferior to those of the business community. Their failure to take an absolutist position in rule making, they fear, will allow their opponents to overwhelm them with one-sided information. Environmental groups that are convinced they are both seriously outnumbered and (to use Michael Pertschuk's phrase) "on the side of the angels" will tend to use every available political resource. The leaders of environmental groups, after all, are advocates. They push as hard as they can because they know their opponents will do the same.

Third, for environmental groups the indirect consequences of some rules are more important than their direct effects. For example, environmentalists pushed transportation control plans in the early 1970s because they wanted to restructure the transportation systems of major cities, not because they believed that ozone constituted a monumental health threat. Similarly, they viewed EPA's prevention of significant deterioration (PSD) regulations as a way to regulate land use and not just pollution (Melnick, 1983:Chaps. 4 and 9). The political allies of environmental groups also have hidden agendas. Eastern coal producers backed the "percentage reduction" requirement for coal-burning power plants for protectionist reasons (Ackerman and Hassler, 1982). According to Bernard Frieden (1979:5), in California, "[r]esistance to growth. . .turned into general hostility toward homebuilding for the average family, using the rhetoric of environmental protection in order to look after the narrow interests of people who got to the suburbs first." The rhetoric of environmental protection—especially when it is freed from the need to answer such questions as how much something will cost and who will pay for it—can serve many masters.

Finally, environmentalists seek not just to lower pollution levels but to raise public consciousness. Once the moral juices are drained from the debate, this job becomes impossible to perform. Similar concerns lie behind labor unions' insistence on strict occupational and health rules. According to John Mendeloff,

> [t]he frequency with which health and safety topics are discussed in union newspapers suggests that they are good political issues for union leaders. More than most issues, they help mobilize a sense of class conflict—of "us" against "them." For this purpose, it helps to draw the lines sharply: unions want the "lowest feasible limit" while the companies want to sacrifice lives for profit. (p. 160)

For some members of the environmental movement, raising public consciousness also means calling into question existing political and economic structures. For them, benefit-cost analysis "legitimizes" not only a level of pollution but also the profit-making system that produces it.

The number of people who consider themselves environmentalists is quite large, and those who are active in environmental organizations are a varied lot. The environmental "movement" ranges from traditional conservationists to the radical "sectarians" described by Douglas and Wildavsky (1982). The issue of benefit-cost analysis may eventually separate those whose chief interests are health, safety, and prudent use of natural resources from those with a much broader political agenda.

CONCLUSION

No one opposes environmental protection per se. Few sane people enjoy pollution or despise scenic vistas. Environmental protection seldom raises troublesome racial issues, and it does not divide people sharply along class lines. The real political issue is always that of opportunity costs: What is given up in reducing water pollution or protecting the snail darter or creating a national wilderness area? If environmental benefits were costless, regulation would generate virtually no controversy.

It took a long time in the United States for environmental issues to reach the national agenda. Environmental regulation was viewed either as improper interference with private property or as the bailiwick of state and local governments. The federal government's position was similar to that of a parent dealing with a rebellious teenager: "I don't even want to talk about it." Matthew Crenson (1971) has referred to this phase as "the unpolitics of pollution."

For reasons that are not yet entirely clear, environmental protection suddenly burst upon the national scene in 1969–1970. At this point the United States' complex system of "separated institutions sharing power"—a system that had previously inhibited action by the federal government—created a bias in favor of stringent regulation. Why? This paper has suggested that the explanation lies in the fact that the structure of U.S. governmental institutions makes it relatively easy for many actors to ignore the only rationale for limiting efforts to protect the environment, namely, opportunity costs. Both Congress and the courts have taken strong—indeed, utopian—positions and delegated to others the job of clarifying and imposing the concomitant costs. Confronted with these legislative and judicial demands, even the most conscientious administrators have taken actions they consider extremely unwise. (The classic example is the transportation control plan EPA announced for Los Angeles in 1973. Referring to the fact that he acted under court order, EPA Administrator William Ruckelshaus joked, "Faced with a choice between my freedom and your mobility, my freedom wins.") Stringent, often unattainable standards provide political benefits for several groups: congressmen who wish to embarrass and berate the executive branch; Democrats seeking to show that Republican presidents have no respect for the environment or for human life; environmentalists who want to keep industry constantly on the defensive and in ill repute.

This is not to say, however, that our political system ignores the cost of environmental regulation. A few laws specifically mandate the balancing of benefits and costs. EPA has on occasion moved toward an explicit comparison of costs and risks. Still, the most common techniques for lowering regulatory demands are "feasibility" requirements and the use of enforcement discretion. These safety valves eliminate the most visible, most politically damaging forms of economic cost: plant closings and layoffs. Another technique is described by John Mendeloff (1986 and 1987); that is, refusing to admit that a substance is potentially dangerous because the regulatory consequences of making this admission are so draconian. As many commentators have pointed out, each of these political coping mechanisms generates significant inefficiencies (Ackerman and Hassler, 1981; Harrison and Portney, 1981; Lave and Omenn, 1981; Crandall, 1983).

Ironically, hostility to the use of benefit-cost analysis may do more to inhibit the quantification and comparison of regulatory benefits than it does to inhibit the consideration of economic costs. As former EPA official Albert Nichols pointed out at the conference, "[w]hat was regarded as illegitimate and regarded with great suspicion was. . .trying to quantify the physical benefits, particularly if one were dealing with non-carcinogens" (conference transcript:128). It is quite likely that the current helter-skelter approach has led us to focus too heavily on certain types of health risks—especially cancer—and consequently to ignore others. As Nichols also stated,

> [i]f you don't have that kind of discipline in the system, there is a tendency to just make qualitative statements which don't allow you to set priorities and don't allow you to deal with the most serious environmental problems. So, we end up diddling away our time with things like Section 112 of the Clean Air Act which involve perhaps dozens of cancer cases a year as opposed to the big hitters like chlorofluorocarbons. (conference transcript:130)

In short, without quantitative evidence, it is difficult to set reasonable environmental priorities.

Blame avoidance is contagious: agency officials frequently question why they should admit that some risks are acceptable when no one else will. Yet responsibility may prove contagious as well. If administrators (preferably those in regulatory agencies rather than in OMB) are forthright and explicit about the need to balance costs and risks and if the courts give them sufficient leeway (as the D.C. Circuit and the Supreme Court now seem to be doing), then the onus will be on Congress to provide more precise and honest statutory guidance. This situation was what occurred with deregulation of the airlines, the trucking industry, and telecommunications. Administrators acted first, the courts deferred, and Congress was forced to decide whether to defend regulatory regimes that benefited only a small group of producers and unions. The status quo crumbled with remarkable swiftness (Derthick and Quirk, 1985). The same process may be occurring with regard to hazardous air pollutants. Once the courts accepted EPA's policy of balancing costs and risks, the burden developed on Congressman Waxman and his allies to garner support for a tougher alternative. So far, Congress has taken no action.

Environmental advocates in Congress and in environmental organizations should view these developments not as defeats but as opportunities. As environmental protection programs grow in number and complexity, it is important to weed out those that focus on lesser problems in order to make scarce resources—expertise, agency money, public support, corporate investments—available for more important programs. This approach will make regulatory policy less of a morality play but more successful in protecting the environment.

REFERENCES

Aberbach, Joel
    1987   The congressional committee intelligence system. *Congress and the Presidency.*
Ackerman, Bruce, and William Hassler
    1981   *Clean Coal/Dirty Air.* New Haven, Conn.: Yale University Press.
Badaracco, Joseph, Jr.
    1985   *Loading the Dice: A Five Country Study of Vinyl Chloride Regulation.* Boston: Harvard Business School Press.
Bardach, Eugene, and Robert Kagan
    1982   *Going By the Book: The Problem of Regulatory Unreasonableness.* Philadelphia: Temple University Press.
Brickman, Ronald, Sheila Jasanoff, and Thomas Ilgen
    1985   *Controlling Chemicals: The Politics of Regulation in Europe and the United States.* Ithaca, N.Y.: Cornell University Press.
Calabresi, Guido, and Philip Bobbitt
    1978   *Tragic Choices.* New York: Norton.
Coppock, Rob
    1985   Interaction between scientists and public officials: A comparison of the use of science in regulatory programs in the United States and West Germany. *Policy Sciences* 18:371.
Crandall, Robert
    1983   *Controlling Industrial Pollution: The Economics and Politics of Clean Air.* Washington, D.C.: Brookings Institution.
Crenson, Matthew
    1971   *The Un-Politics of Air Pollution: A Study of Non-Decisionmaking in the Cities.* Baltimore, Md.: Johns Hopkins Press.
Derthick, Martha, and Paul Quirk
    1985   *The Politics of Deregulation.* Washington, D.C.: Brookings Institution.
Diver, Colin
    1981   Policymaking paradigms in administrative law. *Harvard Law Review* 95:393-434.
Douglas, Mary, and Aaron Wildavsky
    1982   *Risk and Culture.* Berkeley: University of California Press.
Fiorina, Morris
    1977   *Congress: Keystone of the Washington Establishment.* New Haven, Conn.: Yale University Press.
Florio, James
    1986   Congress as reluctant regulator. *Yale Journal on Regulation* 3:351-382.
Frieden, Bernard
    1979   *The Environmental Protection Hustle.* Cambridge, Mass.: MIT Press.
Garland, Merrick
    1985   Deregulation and judicial review. *Harvard Law Review* 98:505-591.
Harrison, David, and Paul Portney
    1981   Making ready for the Clean Air Act. *Regulation* March/April:24-31.

Landy, Marc
1986   Cleaning up Superfund. *The Public Interest* 85:58.
Lave, Lester, and Gilbert Omenn
1981   *Clearing the Air: Reforming the Clean Air Act.* Washington, D.C.: Brookings Institution.
Lipset, Seymour Martin
1986   Beyond 1984: The anomalies of American politics. *PS* 19:222-226.
Lipset, Seymour Martin, and William Schneider
1987   The confidence gap revisited. *Political Science Quarterly* 102:3.
Litan, Robert, and William Nordhaus
1983   *Reforming Federal Regulation.* New Haven, Conn.: Yale University Press.
Mashaw, Jerry, and David Harfst
1987   Regulation and legal culture: The case of motor vehicle safety. *Yale Journal on Regulation* 4:257-316.
McCubbins, Mathew, Roger Noll, and Barry Weingast
1987   Administrative Procedures as Instruments of Political Control. Working Paper no. 109. Center for the Study of American Business, Washington University.
Melnick, R. Shep
1983   *Regulation and the Courts: The Case of the Clean Air Act.* Washington, D.C.: Brookings Institution.
1984   Pollution deadlines and the coalition for failure. *The Public Interest* 75:123.
Mendeloff, John
1986   Regulatory reform and OSHA policy. *Journal of Policy Analysis and Management* 5:440-468.
1988   *The Dilemma of Toxic Substance Regulation.* Cambridge, Mass.: MIT Press.
Mitchell, Robert Cameron
1984   Public opinion and environmental policy in the 1970s and 1980s. Pp. 51–74 in Norman Vig and Michael Kraft, eds., *Environmental Policy in the 1980s: Reagan's New Agenda.* Washington, D.C.: CQ Press.
Moe, Terry
1985   Control and feedback in economic regulation: The case of the NLRB. *American Political Science Review* 79:1094-1117.
National Commission on Air Quality
1981   *To Breathe Clean Air.* Washington, D.C.: Government Printing Office.
Pierce, Richard J.
1988   Two problems in administrative law: Political polarity on the District of Columbia circuit and judicial deterrence of agency rulemaking. *Duke Law Journal* 300-328.
Rhoads, Steven
1985   *The Economist's View of the World: Government, Markets, and Public Policy.* Cambridge: Cambridge University Press.
Rothman, Stanley
1979   The mass media in post-industrial society. Pp. 346-388 in S.M. Lipset, ed., *The Third Century: America as a Post-Industrial Society.* Stanford, Calif.: Hoover Institute Press.
Rothman, Stanley, and S. Robert Lichter
1987   Elite ideology and risk perception in nuclear energy policy. *American Political Science Review* 81:383–404.
Scalia, Antonin
1978   Vermont Yankee: The APA, the D.C. Circuit, and the Supreme Court. *Supreme Court Review* 1978:345.

Schultze, Charles
   1977   *The Public Use of Private Interest*. Washington, D.C.: Brookings Institution.
Shepsle, Kenneth
   1978   *The Giant Jigsaw Puzzle: Democratic Committee Assignments in the Modern House*. Chicago: University of Chicago Press.
Shapiro, Martin
   1986   APA: Past, present, future. *Virginia Law Review* 72:447–492.
Stewart, Richard
   1975   The reformation of American administrative law. *Harvard Law Review* 88:1667.
Taylor, Serge
   1984   *Making Bureaucracies Think: The Environmental Impact Statement Strategy of Administrative Reform*. Stanford, Calif.: Stanford University Press.
Tolchin, Susan, and Martin Tolchin
   1983   *Dismantling America: The Rush to Deregulate*. Boston: Houghton Mifflin.
U.S. Congress, House
   1981   *Role of OMB in Regulation*. Subcommittee on Oversight and Investigations, Committee on Energy and Commerce. 97th Cong., 1st sess. Washington, D.C.: Government Printing Office.
   1985   *EPA's Asbestos Regulation: Report on a Case Study on OMB Interference in Agency Rulemaking*. Subcommittee on Oversight and Investigations, Committee on Energy and Commerce. 99th Cong., 1st sess. Washington, D.C.: Government Printing Office.
Vogel, David
   1986   *National Styles of Regulation: Environmental Policy in Great Britain and the United States*. Ithaca, N.Y.: Cornell University Press.
Wald, Patricia
   1986   The realpolitik of judicial review in a deregulation era. *Journal of Policy Analysis and Management* 5:535-546.
Weaver, R. Kent
   1987   The politics of blame. *The Brookings Review* 5:43.
Weingast, Barry, and Mark Moran
   1982   The myth of runaway bureaucracy—the case of the FTC. *Regulation* May/June: 33–38.

4

Benefit-Cost Analysis as a Source of Information About Welfare

PETER RAILTON

"Benefit-Cost Analysis: Threat or Menace?" is the theme of many philosophical discussions, but it is not the theme of this one. It does not seem to me that the very idea of benefit-cost analysis is a moral outrage or a conceptual absurdity. On the contrary: I am prepared to think that good benefit-cost analysis could do much to improve the reasonableness of policy making. At the same time, however, I am quite uncertain as to whether the actual application of benefit-cost analysis has generally had a salutary effect on environmental policy making, in part because I have some doubts about the justifiability of certain practices that appear common in actual applications of the process. My aim in this paper is to make some of these doubts clear and to indicate roughly how benefit-cost analysis might be carried out in a way that would substantially mitigate them.

My strategy in discussing these doubts is inspired by a question put to me by someone involved in the practice of benefit-cost analysis: "What do I have to believe in order to do this?" I will consider various actual and possible practices in benefit-cost analysis as it is applied to governmental regulation of risk and discuss what one might have to believe in order to defend such practices. I will then ask whether such beliefs seem tenable. When they do not, I will suggest what one might more credibly believe about the matter at issue and how these changes in belief might affect the practice of benefit-cost analysis.

The first part of this paper considers several questions about the scope and comprehensiveness of benefit-cost analysis in policy evaluation. In particular it is concerned with the compatibility of benefit-cost analysis with conceptions of social justice in which distribution, desert, or entitlement

Peter Railton is associate professor in the Department of Philosophy at the University of Michigan.

play a role. The second part of the paper deals with the notion of utility that might lie behind the economist's appeal to preferences in social policy assessment, and the relation of this underlying notion to those of price and willingness to pay. Of special interest is how diminishing marginal utility affects the use of prices to measure willingness to pay. The third part of the paper looks at the question of measuring costs and benefits that will arise in the future. This section argues the need for a distinction between, on the one hand, discounting future prices or willingness-to-pay indicators, a procedure that may be necessary for the accurate measurement of future costs and benefits, and, on the other hand, discounting future utilities themselves, a procedure that may lead to systematic mismeasurement. By way of conclusion, the paper suggests a role for benefit-cost analysis in highlighting future utilities that would otherwise lack adequate representation in the present. In contrast to its reputation as a threat to intergenerational justice, benefit-cost analysis could—in virtue of its commitment to accuracy in measurement—provide information that would make it uncomfortable for policy makers to ignore intergenerational inequities.

I have been asked to contribute a philosopher's perspective to the assessment of benefit-cost analysis. My hope is that this paper engages constructively with the practitioner's concerns in raising the question, "What do I have to believe?" I should, however, emphasize at the outset that my suggestions come from the standpoint of an outsider to the theory and practice of benefit-cost analysis, without the advantages—and disadvantages—of close involvement in the state of the art.

I should emphasize, too, that I am an outsider to the politics of environmental regulation. In particular, I am unable to address the question of whether benefit-cost analysis is a Trojan horse for a particular political tendency in environmental policy. If that currently is where the real issue over the utilization of benefit-cost analysis lies, then this paper is an exercise in naivete and I might be compared to a loyal but benighted citizen of Troy who offers the Greeks advice on how best to paint their gift.

Finally, I should also emphasize that I make no attempt to survey the full range of moral and philosophical issues that surround the justification of benefit-cost analysis, but instead focus almost entirely upon certain questions related to the measurement of benefits and costs. I have chosen this focus for a particular reason: if benefit-cost analysis fails in its job of measurement, then its greatest claim to the attention of those involved in environmental regulation—its capacity to provide information about the relative magnitude of a wide range of disparate benefits and costs, some of which might otherwise receive little notice or quite arbitrary treatment—will become too weak to justify consideration of its use. As I hope to indicate, however, benefit-cost analysis may be able to do a reasonably good job of measurement if those applying it are explicit about the difficulties involved and if they adjust their claims about the decisiveness, accuracy, and comprehensiveness of its assessments to the reality of its limited measurement techniques.

SCOPE AND COMPREHENSIVENESS

To carry out benefit-cost analysis, what must an individual believe about its scope? The short answer to this question is, "Very little, if one puts it only to very little use." In contrast, if a person thinks benefit-cost analysis can be a powerful tool of policy assessment, then he or she must have fairly powerful beliefs about what can come within its scope.

Let us consider, for example, an extreme position that is probably held by very few people: benefit-cost analysis affords the appropriate normative criterion for social choice. To believe this, one would have to believe that all the benefits and costs relevant to the assessment of policies can be satisfactorily accommodated within benefit-cost analysis.

How hard would it be to believe that? Benefit-cost analysis is founded on the idea of a potential Pareto improvement, that is, a change in which the gainers benefit sufficiently to be able to compensate the losers and still come out ahead. How might this standard—which is usually seen as capturing an idea of efficiency—provide the appropriate normative criterion for social choice? The clearest rationale would probably run as follows. Changes meeting this standard can be thought of as enlarging the total pool of benefits that are socially available. If certain complications are ignored and if population remains constant, and moreover if it is assumed that over time the benefits and burdens of social policies tend to be distributed randomly among individuals, then, when policies are implemented that all satisfy a standard of potential Pareto improvement, the expected value of all social positions will tend to rise. This rationale comes close to capturing a classical utilitarian standard of social choice (because overall utility increases if the expected value of all social positions rises). More surprisingly, it comes close to capturing a hypothetical contract standard of social choice because rational individuals who are concerned about advancing their own well-being will tend to prefer social arrangements with higher expected utility per position. Because hypothetical contract theory and utilitarianism are the two dominant philosophical trends in the theory of social justice, it might be argued that reliance on a potential Pareto improvement standard would be consistent with promoting justice in well-recognized senses.

Yet this argument does not really make it easy to believe that benefit-cost analysis gives an exhaustive account of the values that are relevant to normative social choice—even the single value of justice. In the first place, this approach assumes that benefits and burdens are randomly distributed; in the absence of further argument, however, it is hard to believe that this condition will be met so reliably that distribution in particular cases or in cumulative patterns need not be considered. After all, the standard of potential Pareto improvement says nothing about existing inequalities, and there is no guarantee that changes that would meet the standard would not bring about arbitrarily large increases in inequality. Serious normative worries might therefore arise for distribution-sensitive conceptions of justice.[1] In the second place, the standard says nothing about whether the present distribution is in any sense deserved, and there is no guarantee that changes that would meet the standard would not allow gains or losses to go to individuals who did not earn or otherwise deserve them. Serious normative worries might therefore arise for desert-based conceptions of justice. Even if one were to suppose that many potential Pareto improvements would be compatible with desert or equity, some would not be. Thus, on the basis of most conceptions of justice, the benefit-cost standard could not be an adequate criterion of normative social choice.

It seems to me that most theorists and practitioners of benefit-cost analysis recognize these limitations and do not advocate the use of benefit-cost analysis as a *sufficient* condition for policy approval. It is not uncommon, however, to hear benefit-cost analysis put forward as a *necessary* condition for approval, a preliminary "test" that policies must meet before further tests (of equity, desert, and so on) can be applied. Yet reasonable demands of equity or desert may sometimes require changes of a kind that would not pass a potential Pareto improvement test. For example, redistribution designed to remedy past injustices might not yield more for the gainers than it costs the losers; still, the gainers might deserve the change, and equity might be served by it. In a society in which past injustices call for such remedies, to restrict social choice to policies that represent potential Pareto improvements would be to decide, in effect, against these demands of desert or equity.

A defender of benefit-cost analysis might respond to these considerations by suggesting that the scope of benefit-cost analysis be enlarged to encompass all normatively relevant features in social choice. One way to accomplish this would be to identify people's willingness to pay for equity or desert. Yet surely, the question policy makers face is not simply "How much do our constituents care about whether things are just by their own lights?" but also "What *is* justice—what does it require?" However useful it may be to know the answer to the former question in order to answer the latter, the former cannot entirely replace the latter. In many cases, considerations of justice appear to function not as factors or weights within

---

[1] Such conceptions include not only hypothetical social contract theory of the Rawlsian variety (see Rawls, 1971) but also classical utilitarianism, which, owing to diminishing marginal utility, is also distribution sensitive. This question is discussed in the second section of this paper.

benefit-cost analysis but rather as constraints on the proposals that are prepared for such analysis. A benefit-cost analysis of the regulation of a toxin does not even consider such cost-cutting options as massive government deception about health effects or refusing to offer health care to individuals of certain races. A utilitarian might argue that such alternatives need not be analyzed because people are already convinced that they are contrary to maximizing social benefit. A Kantian, on the other hand, might insist that, regardless of whether social benefit is maximized, the government should not deceive or discriminate in these ways. These viewpoints constitute two poles of a substantive debate about the foundations and requirements of justice. Fortunately, it is not necessary to resolve this debate, or even enter into it, in order to carry out benefit-cost analysis—that is, as long as a benefit-cost standard is not treated as the appropriate normative criterion of social choice.[2]

If a benefit-cost standard is not seen as a necessary or sufficient condition of social choice, how, then, should benefit-cost analysis be viewed? It might be used as an information-yielding device, a way of generating and bringing together within a quantified scheme a great deal of data about how people are likely to be affected by alternative policy choices. Both utilitarians and contractarians consider it important to know how policies compare with regard to the sorts of benefits that benefit-cost analysis is best equipped to measure. For example, if a policy appears to be consistent with the constraints of justice and yet is found to be inefficient, or not cost-effective, or suboptimal with respect to aggregate expected value, that information will be viewed as normatively relevant by utilitarians and contractarians alike. One need not take sides in a controversy over the theory of justice to assign considerable importance to the information yielded by benefit-cost analysis in social decision making.

Viewing benefit-cost analysis as an information-yielding rather than a decision-making device has implications for how such analyses should be presented. For example, a benefit-cost analysis will be more informative (although less decisive) if it reports disaggregated as well as aggregated effects—for instance, by telling policy makers which elements within the population will receive which costs and which benefits. If benefit-cost analysis is to help policy makers to choose justly, and if justice is distribution- or desert-sensitive, then disaggregated information may at times be more crucial than overall net results. Similarly, a benefit-cost analysis will be more informative (although less conclusive) if it refrains from collapsing

---

[2] Some approaches (e.g., Ben-David, Kneese, and Schulze, 1979) have sought to capture these competing conceptions of justice within something like a benefit-cost framework. These attempts come to grief, however, in their inability to achieve an adequate representation of a fundamentally deontological theory of justice. The agent-centered constraints of such theories may simply lack a non-agent-centered, or global, value-theoretic expression.

all uncertainties and reports error bands or sensitivities rather than point measures. There are risks to justice that are every bit as significant as risks to health; it is surely a matter of justice how much of such risk governments should permit. It may be tempting to practitioners of benefit-cost analysis to see their real goal as a grand sum; it may also be tempting to policy makers to say to analysts, "Look, just tell me: what's the bottom line?"—perhaps in the hope that this figure will somehow decide difficult questions of justice for them. Such temptations should be resisted, however, if benefit-cost analysis is to play its appropriate role in policy deliberations.

Thus far, the discussion has centered on possible justice-based constraints on benefit-cost analysis without singling out any particular conception of justice. Such generality seems appropriate: because there is genuine controversy over the nature of justice and what it requires, it would be inappropriate to consign to an analytic technique the task of deciding which social outcomes are to be sought in this country.[3]

Surprisingly, the range of conceptions of justice that would accord a significant role to the sort of information produced by benefit-cost analysis may be even broader than has thus far been suggested. There may be such a role even within the conception of justice that seems most opposed to social aggregation and balancing: Lockean natural rights theories (e.g., Novick, 1974). According to the more extreme forms of such theories, individual rights in person and property are natural rights of exclusion existing apart from social arrangements and constraining permissible social policy by setting up normative boundaries that cannot be crossed without the consent of the owner. Thus, for example, if the government were to allow dust and diesel fumes from the building and operation of a new intercity rail line to drift, unbidden, onto private property, then the integrity of that property would have been violated. It would be no defense, in this view, for the government to argue that the aggregate benefit realized from the new rail line would exceed the aggregate cost it would impose. A Lockean natural rights conception of justice, is disaggregative, since each individual is entitled to say what can and cannot cross his or her property line. In the example, if a given individual does not want the benefits or burdens of the new line, then the government is not free to send its dust and fumes onto his or her property, however beneficial the line might be to society as a whole. If control of all dust and fumes would be prohibitively expensive, then private property rights may stand in the way of building the line at all.

---

[3] Even a utilitarian might justifiably believe that the proper role for benefit-cost analysis is in providing information rather than in decision making because a great many factors influencing direct or indirect utilitarian assessment are not well captured by the specific analytic techniques of benefit-cost analysis.

Of course, were the government to secure the voluntary consent of all those whose property boundaries would be at risk of violation, construction could proceed. Given the nastiness of dust and diesel fumes, individuals presumably would demand some compensation in return for their permission. With person and property at stake, individuals would be entitled, on a natural rights theory, to hold out for whatever they could get—even to refuse to consent at any price. As a result, a single individual could wield veto power over a scheme of wide-ranging social benefit, and this would give rise to perverse incentives. Individuals could wait until the government obtained near-unanimous consent to some beneficial scheme, and then hold out for compensation much in excess of any harm they might themselves bear.

Yet the constraints of individual veto power and its perverse incentives might be removed if it were possible to go ahead with the beneficial scheme and then provide after-the-fact compensation to all whose property had been adversely affected. This way of proceeding would not allow individuals to set their own levels of after-the-fact compensation, lest they demand arbitrarily large sums and in that way continue to exercise an effective veto. Instead, some fairly objective measurement of the extent of harm to individuals would be established so that commensurate compensation could be determined. The test of potential Pareto improvement addresses just this sort of measurement problem, although its application within a Lockean scheme would have to involve actual rather than merely possible compensation. This requirement in turn would mean including transaction costs in the assessment. Once that were done, however, a policy or activity that appeared to be a potential Pareto improvement could be pursued under this violate-and-compensate gambit.[4] Thus, a government—or, for that matter, individuals—would be greatly interested in receiving the sort of information that benefit-cost analysis could provide using a framework of actual compensation.

From a philosophical point of view, it is not clear whether those whose fundamental sympathies are Lockean should be happy with the violate-and-compensate gambit, for it may be incompatible with full respect for individual property and consent. Yet a Lockean scheme without this gambit might result in so many inflexibilities as to disqualify it for any actual application. In the modern interconnected world, it is simply not possible to secure actual consent to all impositions of risk or to refrain from all risk-imposing activities. Indeed, even with the gambit, the Lockean scheme would be extraordinarily cumbersome, owing to the problem of identifying

---

[4] An example of this approach is eminent domain. It prevents individuals from exercising veto power over public projects or extracting exorbitant sums in exchange for giving their consent, and it involves compensation by some objective standard.

and appropriately compensating all actual victims.[5] For now, however, it is sufficient to note that Lockeans who are concerned with real-world applications are bound to seek something like this sort of flexibility and that for such purposes it would be important to have the sort of information about relative magnitudes of harm and good that benefit-cost analysis may afford.

In sum, individuals with various perspectives on justice may find the information produced by benefit-cost analysis to be greatly valuable. Once benefit-cost analysis is understood as a process meant to yield information rather than to make decisions, practitioners of benefit-cost analysis need not take sides in controversies over the nature of justice. At the same time, it would be quite controversial—indeed, it would almost certainly amount to taking sides against widely held deontological conceptions of justice—to attempt to assimilate the theory of rights into a benefit-cost framework by assigning monetized costs to violations of rights and then entering such costs into an aggregative, balancing scheme. In a range of cases, it would seem that rights are better understood either as (a) constraints upon the possible projects whose costs and benefits decision makers are prepared to assess by analytic means or, more weakly, as (b) markers of areas of special social concern in which policy makers are uncomfortable with a straightforward balancing of costs and benefits. Either way, decision makers will want to give the issue of rights a distinctive role in their deliberations and not blend such considerations into a homogeneous mixture of costs and benefits.[6]

MEASURING COSTS AND BENEFITS
AT PARTICULAR POINTS IN TIME

What does an individual have to believe he or she is measuring when performing benefit-cost analysis? I have been pursuing the idea that benefit-cost analysis should be seen as an information-yielding process. What, then, does benefit-cost analysis provide information about, and how accurate can it be?

One view on this matter has already been called into question, namely, the view that benefit-cost analysis provides an assessment of the relative normative weight of all considerations, including all matters of justice or rights. It seems to me much less implausible to believe that benefit-cost

[5] Nozick (1974) considered adding such a gambit to Lockean theory, although he also showed some preference for using a system of tort law to carry out compensation. For further discussion of risk and Lockean theory—with or without the gambit—see Railton (1986b).

[6] I should perhaps emphasize that this claim does not depend on rejecting a utilitarian analysis of rights. It is an open question whether the system of rights considered optimal by a utilitarian would permit various sorts of balancing in public policy making. The optimal system of rights in the long run might be one that imposed some fairly strict limits on balancing.

analysis finds its proper subject matter in the realm of welfare. This view might seem to run counter to the orthodox conception that benefit-cost analysis is based solely on the notion of preference. Yet why, when social policy is being chosen, do decision makers think that costs and benefits based on individual preference have any special relevance? Why not look to some less "subjective" measure? The usual answer involves the principle of consumer sovereignty: other things being equal, policy makers should respect the individual's judgments about what matters to him or her and how much it matters. This principle is not a piece of democratic theory, to the effect that "the people shall judge." For example, the principle of consumer sovereignty is applied to preferences but not to beliefs. Benefit-cost analysts seek out expert opinion on the likely effects of risky activities on the environment or on human health; they feel no compulsion to base their estimates of the probability of outcomes on some average drawn from popular opinion. Why, then, do they seek to base their estimates of the utilities of outcomes on individual judgments?

The answer presumably is that, although the public believes that scientists are more knowledgeable than the average American about natural phenomena, it does not believe that, in general, a group of experts would be more reliable than any given individual in ascertaining the extent or nature of that individual's welfare—not, at least, once that individual was informed about the options he or she faced. Scientists have the most sustained and most detailed experience of the natural world, but individuals have the most sustained and most detailed experience of how choice and outcomes affect their particular well-being. This answer would also explain why the principle of consumer sovereignty is applied to the preferences of adults rather than children. If individuals learn about the nature of their well-being through experience, and especially the experience of choice, and if moreover adults in general have better information about the options they face than do children, then the opinions of adults are more likely to be accurate.

By contrast, if one were simply an outright skeptic about the notion of welfare or about the possibility of learning about one's own good, there would be no clear rationale for giving more credence to the opinion of a given person than to the opinion of some randomly selected third party about the effects of a given social policy on that individual's well-being. In addition, there would be no clear rationale for granting informed preferences more authority than uninformed ones. The most plausible way to provide such rationales, and thereby to underwrite the theory and practice of benefit-cost analysis, is to believe that there is such a thing as individual welfare and that individuals do a better job of recognizing it in their own case than do third parties, especially when individuals are well informed.[7]

It may be somewhat unfashionable to speak in such terms. At times, it seems that, some contemporary economists wish to see no more in the notion of preference than the mere fact of consumer choice among bundles of goods. Yet unless one believes that behind choice lies an effort by individuals actually to achieve their goals, it is unclear why choice behavior even warrants the term *preference*. It certainly becomes problematic to determine how choice behavior could be a basis for deciding whether and to what extent individuals have received benefits, borne costs, or undergone compensation. What relevance would benefit-cost analysis have in normative policy making unless it was believed that preference and willingness to pay tended to reflect real gains or losses to the quality of individual lives?[8]

Perhaps the concept of well-being has been out of favor in economics circles because well-being, unlike behavior, is not publicly observable. Yet there is nothing inherently unscientific in positing the existence of unobservables—whether electrons or viruses or utilities—to organize and explain observation, as long as one is also responsive to evidence. Indeed, there is something especially odd about calling well-being "unobservable"; unlike an electron, it is something of which each person has had the most intimate experience. Furthermore, given the similarities among all human beings (e.g., they are all made of flesh and blood), it seems idle to imagine vast differences in the range and character of the inner lives of those whose environment and behavior are broadly similar. Indeed, this worry is really closer to metaphysical skepticism about other minds than to the methodological concerns of scientific psychology.[9]

---

[7] In their defense of benefit-cost analysis, Leonard and Zeckhauser (1986:35) put forward "the basic tenet that people know what is good for them," based on "the assumption that agents are well informed." For a philosophical defense of a conception of individual good based on something similar to informed preference, see Railton (1986a).

[8] The question of factors other than individual welfare that influence preferences (e.g., social values, posthumous goals) is considered later in this paper.

[9] These remarks suggest a reply to die-hard positivists. Even classical behaviorism has availed itself of the notions of positive and negative reinforcement, aversion, satiation, expectation, and so forth, by giving them operational definitions. At least one strain of revealed preference theory seems to treat preference in a similarly operationalist way. Questions about whether a given choice *really* reveals an individual's preferences are blocked by pointing out that this definition simply follows from the stipulation of what "revealed preference" is to mean. Is there any reason, then, why the positivist should not consider operationalizing the notion of cardinal utility by tying it to specific patterns of behavior—perhaps along with physiological evidence concerning, say, galvanic skin response and pupil dilation? If the objection is raised that it is difficult to see the interest, for purposes of social choice, in working with such a concept of utility unless it

Let us say, then, that the best case to be made on behalf of the relevance of benefit-cost analysis to normative social choice is that it provides systematic information concerning the extent to which a person's well-being will be affected by alternative policies. In principle, benefit-cost analysis is able to provide this sort of information because it attaches monetized values to the benefits and costs experienced by different individuals and therefore permits comparison and aggregation. In this respect, it resembles a Benthamic felicific calculus. The mechanism for the assignment and comparison of values differs from Bentham's, however, because the test of a potential Pareto improvement appears to involve no appeal to interpersonal comparisons of cardinal utility.[10] Let us suppose, for example, that Mr. Smith could receive benefit B only if risk R were imposed on Mr. Jones. Yet suppose as well that Jones is indifferent regarding the choice between continuing the status quo and bearing additional risk R while receiving compensation $\$C$; suppose, too, that Smith is indifferent regarding the choice between continuing the status quo and receiving benefit B at a cost to him of $\$D$. If $\$D$ is greater than $\$C$, imposing risk R to bring about benefit B passes the potential Pareto improvement test. Smith gains enough to compensate Jones—although it is not assumed that he actually does so—and still come out ahead.

It should be noted, however, that the choice behavior just mentioned was described in the language of *indifference, benefit, cost,* and *compensation.* Were one to discover that the behavior of Jones, which underlies the imputation of indifference, was the result of real or imagined intimidation by Smith or, alternatively, of a simple misunderstanding of what was being asked of him, then his behavior would not support an interpretation in such terms (indifference, compensation, etc.). Moreover, if these terms did not fit, it would be difficult to believe that the test revealed anything interesting about gains or losses in well-being. To take the test of potential Pareto improvement seriously for purposes of social choice, it is necessary to believe that the sorts of choice behavior on which the assessments are based can be plausibly interpreted as reflecting the choosers' views about how their well-being is likely to be affected.

Now, the real question of this section can be posed: Under what circumstances will a test of potential Pareto improvement reliably be informative about net effects on well-being? Obviously, it is necessary to ensure as far as possible that the parties to the choice behavior in question are neither in reality nor in their imagination being coerced and that they are

---

corresponds to something less stipulative, I will gladly concede the point—at least so long as it is recognized that the objection arises with equal force against revealed preference.

[10] Whether, in a sense, the test actually involves a form of interpersonal comparison is discussed later.

not making simple mistakes of perception. Five other sources of possible inaccuracy may be of special relevance to the case of environmental risk assessment.

Diminishing Marginal Utility in the Intrapersonal Case

I have been suggesting that behind the use of choice behavior and the imputation of preferences lies the view that there really is such a thing as an individual's welfare and that, with experience, individuals tend to acquire some knowledge about their welfare and to act accordingly. One thing that can be said with considerable confidence about individual welfare is that most benefits exhibit diminishing marginal utility, a phenomenon people experience directly in their own lives.

Recognizing this phenomenon has an important effect on the interpretation of tests of potential Pareto improvement. Indicators of willingness to pay that are based on prices or related choice behavior will reflect the marginal utility of "the last unit." For example, an analyst might use the amount I am willing to pay as an entry fee into a wilderness area or for a fishing permit to measure the cost to me of converting some forest area to commercial use or of the environmental degradation caused by acid rain. This measure, however, will only indicate the marginal utility to me of "the last unit" of wilderness consumption, and the policy in question might have as one possible effect a dramatic alteration of the amount of wilderness or fishing available to me. Whenever such nonincremental changes are possible, the proper measure must take into account the higher marginal utility of the "first" unit of consumption, the "second" unit of consumption, and so on. That is, it is necessary to take consumer surplus into account.[11] The issue is not whether dramatic changes are the *likely* result of any given policy but whether they are a *possible* result. If so, then the proper way to calculate the expected value of nonincremental outcomes is to take the product of their probability and the sum of "price" and any consumer surplus.

The discrepancy between this procedure and one based on a marginal measure can be quite large. One need only reflect upon the difference between the amount one would be willing to pay to achieve a 1 percent increase in physical mobility and the amount one would be willing to pay to avoid utter physical immobility—would the former be as much as one one-hundredth of the latter? Yet among the consequences of risky policies are not only marginal effects—the exposure of a population to an incremental amount of unwanted risk—but also, typically, the nonmarginal effect of

---

[11] I am grateful to Hal Varian for pointing out to me that the intrapersonal phenomenon I wish to describe here can be expressed using the economists' notion of consumer surplus.

actual injury to individuals. Here, too, the proper way to calculate the expected value of such nonmarginal outcomes—that is, to reflect the full difference between normal health and significant illness or disability—is to take the product of the probability of the various nonincremental harms and the sum of their "price" plus any consumer surplus.

Diminishing Marginal Utility in the Interpersonal Case

Diminishing marginal utility, I have suggested, is something each person experiences directly. It is, moreover, something each individual exhibits in his or her behavior through changing marginal rates of substitution. One might further say that, in the absence of compelling evidence to the contrary, it is reasonable to assume that people who have similar observable physical and social characteristics and who exhibit similar choice behavior will also be broadly similar in the amount of utility they derive from specific benefits. In particular, it is reasonable to assume that those who are otherwise similar but who stand far apart on a scale of income or wealth will derive different marginal utility from money. For example, certain tasks (e.g., some forms of dirty, difficult, nonself-directed labor) are almost universally disliked. The amount of such activity that can be obtained from individuals simply by holding out the inducement of, let us say, $10, is almost always less when their wealth is great and almost always diminishes as their wealth increases.

Returning to the potential Pareto improvement test as applied to Jones and Smith, let us suppose that Jones is very much like Smith in most respects but is considerably less affluent. For Jones, the marginal utility of a dollar can be expected to be significantly greater than it is for Smith. In the earlier example, Smith would pay $D to receive benefit B and Jones would accept compensation $C to run associated risk R. Yet it does not follow from the fact that $D is greater than $C that the gain to Smith is of sufficient magnitude to offset the loss to Jones because the dollars Smith would pay are paid at a lower marginal utility than the dollars Jones would receive. Of course, if amount $C were actually to be transferred from Smith to Jones as compensation, then no problem would arise because the marginal utility of these dollars would then be that of Jones. It is not part of the potential Pareto improvement test that such a transfer takes place, however. Because there is no actual compensation, the test involves what is in effect an interpersonal utility comparison of a $C-valued-by-Jones loss with a $D-valued-by-Smith gain. Owing to diminishing marginal utility, one would expect that, if such a comparison were based solely on the dollar amounts involved, the measure would be inaccurate in a systematically regressive way, exaggerating gains or losses to the more affluent in comparison with those to the less affluent.

Still, it might rightly be asked whether this type of comparison will lead to systematic inaccuracies in large-scale policy assessment. Under certain conditions, inaccuracies will be reduced: when willingness to pay is assessed using fairly broadly based behavioral indicators (e.g., prices) to obtain average values; and when the benefits and burdens of large-scale policies are rather uniformly dispersed over varied populations. But when such conditions are not met, the test will tend to yield errors. This might not be worrisome if the net errors that remain within assessments of individual policies themselves varied randomly from the assessment of one policy to the next, so that within sequences of policies, the effects of assessment errors would tend to cancel each other out over time. I know of no general reason to expect that this sort of balancing will occur; on the other hand, there is a general argument—based upon diminishing marginal utility— showing a systematic tendency toward nonrandom, regressive error. As a result, it is difficult to believe that policy assessment can legitimately ignore questions about inaccuracies arising from effects of uneven distribution.[12]

For example, errors can creep in if one compares a willingness-to-pay indicator drawn from a population in which the marginal utility of money is relatively high (e.g., a risk premium in the wage of miners) with a willingness-to-pay indicator drawn from a population in which the marginal utility of money is relatively low (e.g., an aesthetic premium in the price of property in settings of unusual natural beauty).[13] A second example involves errors that can arise in comparisons of the benefits of controlling job-site pollutants liable to produce nonincremental health effects in the small exposed population of producers with the cost of incrementally increasing prices to the large population of consumers. A third example combines the intrapersonal and interpersonal effects to magnify the possible error. In this case, a marginal willingness-to-pay indicator is used to measure an incremental cost to one relatively affluent population (e.g., an increase in the price of strawberries); this figure is then set alongside a marginal indicator used to measure a nonincremental harm to a less affluent population (e.g., an increase in disability among migrant laborers who work in strawberry

---

[12] One possible source of balancing is that certain benefits may actually have higher utility for the more affluent, because the enjoyment of such benefits requires the possession of other resources. More benefit from an Alaskan wilderness area may accrue to the more affluent because they, unlike others, are able to get there. It is difficult to believe, however, that this sort of balancing will be sufficient to offset regressive effects in the general run of environmental and health-related cases to which benefit-cost analysis is applied.

[13] A much more important example of this phenomenon, and one with systematic import, concerns future generations—under the assumption that they will be substantially better off than people who are alive today. This question is discussed in the third part of this paper.

fields in which a short-lived but toxic fungicide has been applied).[14] In such cases, some and perhaps considerable adjustment of monetized values would seem to be required to improve the capacity of willingness-to-pay indicators to represent utilities.

However ordinalist its official theory, benefit-cost analysis cannot plausibly dismiss interpersonal comparison and diminishing marginal utility. Even the idea that a potential Pareto test offers a way of measuring increases in the total social pie requires the belief that, for example, Smith's $D-valued-by-Smith gain can be compared with Jones's $C-valued-by-Jones loss. Too much is known about what money, or any other good, means to an individual's well-being to convince anyone that this comparison can be settled simply by asking whether $D is greater than $C. Too much is known, that is, to be persuaded that an uncorrected potential Pareto improvement test captures as fully as possible the welfare effects of social choice.[15]

Preferences Involving Poor Information or Other Cognitive Defects

To the extent that one is prepared to use market prices or choice behavior as indicators of willingness to pay and, ultimately, of welfare, one must believe that the people involved are rational, reasonably well informed, confronted with an appropriate array of options, and so on. Prices, for example, directly measure not willingness to pay but the tendency to pay, a mixture of willingness and real or imagined constraints, of knowledge and ignorance.

In choices in which outcomes (a) depend on complex causal sequences, (b) show great latency, or (c) involve options of markedly different salience, there is good reason to expect everyday choice behavior to embody serious informational and deliberative defects. These conditions notoriously obtain in many areas of choice behavior that are relevant to environmental policy making: labor markets for jobs involving risks, the market for insurance, the

---

[14]This example is not fanciful. It reflects procedures used in an unpublished risk management case study (Conservation Foundation, 1987) that is meant to be fairly sophisticated methodologically. I say "reflects" because the exact bases for the mortality and morbidity values this study uses are not stated, although the values seem consistent with familiar marginal willingness-to-pay indicators.

[15]Corrections to benefit-cost measurement discussed in this section involve attempts to capture welfare and not the problem of capturing other values (e.g., process or equity). The concern with diminishing marginal utility, for example, reflects the need to remove certain distortions of the potential Pareto improvement test or willingness-to-pay indicators. As it happens, these distortions have something like regressive distributive implications. The argument for correcting them, however, does not depend on any assumption favoring the incorporation of some measure of distributive justice into benefit-cost analysis; rather, it depends only on two assumptions: that what is wanted is a measure of welfare effects that is as accurate as possible, and that marginal utility declines.

market for safety devices, and so on. The wages of coal miners tell something about risk premiums, but they also speak volumes about restricted options, historical bargaining positions, and the lack of information.[16] If prices are to be used to estimate willingness to pay, one must either believe that these distortions are not significantly present in the particular markets of concern, or one must attempt to correct prices to reduce the influence of such distortions.

Price correction need not run counter to the fundamental idea underlying consumer sovereignty if the adjustments are based on efforts to create real-life choice situations in which agents are given good and usable information, a wide range of options within which fine discrimination is possible, and ample resources from which to choose. Experiments of this kind, which might bear some resemblance to the famous negative income tax experiments, could be carried out in representative populations on an appropriately large scale. The preferences expressed in such experiments could then be used to develop more general projections concerning how individuals' preferences would evolve were they well informed, well placed to choose, and so on. Consumer sovereignty could thus be upheld, although at the cost of expensive experimentation. It seems to me that, by calling for the funds necessary to underwrite experiments to improve the quality of the information yielded by their work, advocates of benefit-cost analysis could simultaneously exhibit their commitment to empiricism and help mitigate the widespread suspicion that they favor technocracy over autonomy.

Here, too, another worry about benefit-cost analysis might be addressed. It is well known that individuals give markedly different answers to questions about how much they would *pay* to avoid a certain increase in risk, as opposed to how much they would *demand to be paid* to bear an increased risk of the same magnitude. The latter sum tends to be larger than the former, often by several orders of magnitude. Various cognitive mechanisms have been proposed to explain this discrepancy. Pending a better theory of what is going on in such judgments, it seems to me inappropriate—and bound to raise suspicion—that benefit-cost analysis tends to seize on the smaller figure when measuring willingness to pay. No doubt benefit-cost analysts feel that the very large sums demanded for

---

[16]The wages of coal miners may also reveal something about social norms—for example, ideas about what one ought to do to live up to social expectations and to provide for one's family, what deference one owes to those with higher social standing, or what activities are socially valued. Such considerations may in any given case have a significant effect on willingness to pay, although they will greatly complicate the interpretation of choice behavior as a straightforward indication of individual evaluation of risk. A policeman or fisherman may demand less in compensation for a job-related risk that is deemed socially expected or gender appropriate, or that is connected with a socially valued activity, than he would demand for a risk with none of these features (e.g., passive exposure to an environmental toxin). For further discussion of preferences and values, see the next section.

compensation do not reflect the sort of realistic valuation of money that comes into play when individuals are asked what they would be prepared to pay out of pocket. By the same token, however, the rather small sums individuals say they would pay may reflect a lack of realistic understanding of the magnitude of the risks involved (money paid out of pocket being much more salient than probabilities of harm). Accuracy in measurement might be served by working with both figures as something like upper and lower bounds rather than accepting the lower figures as more credible. Both figures seem likely to reflect some departure from cognitively ideal deliberation.

Finally, it is important to take indirect welfare effects into account whenever cognitive defects are present. Let us suppose that a fear based on poor information is widespread and that fuller public information will reduce but not dispel it. A policy that appears to be optimal when its returns are calculated on the basis of informed preferences may nonetheless cause considerable anxiety in suspicious or information-resistant segments of the population. This anxiety and the resultant individual and social dysfunctions are a real welfare cost of the policy and should not be dismissed as unworthy of attention. How much weight to place on unreasonable fears is a substantive matter in the policy arena. The job of benefit-cost analysis, by contrast, is to let policy makers know, as fully as possible, the likely welfare effects of acting in the presence of such fears. In making their assessment, benefit-cost analysts will ensure, to the extent possible, that they have the best available information about the actual harm individuals are likely to experience rather than the harm that uninformed individuals expect to undergo. At the same time, it is important to note that actual harm includes not only direct health effects but also social, psychological, and medical problems arising from (warranted or unwarranted) anxiety.

Preferences Not Related to Welfare

By and large, the preferences that make themselves felt in prices and other indicators of willingness to pay can be expected to reflect individuals' explicit or implicit pursuit of their own well-being. Yet there are various areas in which strongly held values that are largely independent of personal well-being also influence choice behavior.[17] Nowhere is this clearer than in environmental issues, an area in which concerns about posterity, about the kind of society Americans want and the sorts of things they think should be done to bring it about, inextricably enter into people's preferences. This problem may be especially large when political behavior—such as

---

[17] Amartya Sen (1985) especially has drawn attention to such phenomena.

support for various sorts of environmental policies—is used evidence for a willingness to pay.

Typically, as in the case of cognitive defects, there will be indirect welfare effects of such preferences. For example, if I very much want a wildlife area to be protected, then I prefer not to hunt there, and I also prefer that no one else hunt there. Should I learn that people are hunting there, I would be disturbed in a way that affects my well-being. Such indirect welfare effects are a genuine part of any calculation of social welfare, even though their basis lies in other-directed preferences. This approach is to be distinguished from one that gives other-directed preferences a direct role in utility assessment (because arguably, apart from my agitation on learning of hunting in the preserve, my welfare is not affected adversely by the mere fact that hunting is occurring).

If such indirect welfare effects were always of the same magnitude as direct effects on preference fulfillment, it would be possible to ignore the question of whether a given willingness-to-pay indicator reflects self- or other-oriented preferences. It is impossible to believe that these effects will be that similar, however. For example, my willingness to pay for some benefit may reflect preferences I have about how things will go after my death. Once I am gone, however, the extent to which this preference is satisfied can be affected although my welfare cannot.

If the goal in such cases is to assess gains and losses in individual welfare, analysts will be misled to some degree by looking at individual preferences. And yet such a widened scope of interest does not create a fundamental tension with the doctrine of consumer sovereignty. The basis of the doctrine of consumer sovereignty is the individual's reliability to judge his or her own well-being; in the above instance, preferences regarding other matters are being considered. Thus, some adjustment in the interpretation of choice behavior may be called for whenever other-directed preferences can be expected to play a large role.[18] This approach does not at all preclude the measurement of indirect welfare effects, although it requires some attention to the sorts of preferences that are consulted when assessing such effects. Careful analysis would avoid amalgamating other-directed and personal-welfare-based preferences.

---

[18] It may not seem quite so philistine to endeavor to exclude other-directed preferences from direct consideration in benefit-cost analysis once one recognizes the limited, "welfarist" purpose of benefit-cost analysis and once one notices that the values that other-directed preferences reflect quite properly play a significant role in actual policy choice. However, the charge of philistinism may stick to anyone who insists upon treating benefit-cost analysis as the decisive normative criterion of social choice.

The Absence of Appropriate Markets

To believe that benefit-cost analysis can achieve a reasonable degree of accuracy in its information-yielding role, one must believe that the process is able to contend with those benefits that are important to welfare but that have no existing appropriate markets. The welfare of individuals is much affected by many features of the world—for example, the aesthetics of daily life, the degree of community enjoyed—for which it is difficult to produce direct willingness-to-pay measures.

Two approaches come to mind. First, an analyst might give quantitative measures only for those benefits for which an appropriate market exists and note explicitly the incompleteness of the analysis. One way to accomplish this is illustrated by Porter's benefit-cost analysis of mandatory deposits on beverage containers (Porter, 1978). A major benefit that could result from a container deposit law would be a reduction in litter, with attendant aesthetic gains, an increase in community pride, and so on. How is one to give a monetized value to such benefits, however? Porter wisely does not try; instead, he carries out a sensitivity analysis that indicates a range of willingness-to-pay values, within which one's willingness to pay for these difficult-to-quantify goods would have to fall in order to outweigh the estimated costs of the deposit system.

A second approach attempts to find or create some proxy market or other indirect way of assigning a quantitative measure. Per capita government expenditures on litter removal, for example, might be considered as an indicator of willingness to pay. The danger of this approach is its arbitrariness; the danger of the first approach is that nonquantifiable benefits may simply drop out of sight in favor of those that can be more readily measured.

One advantage of the second approach is that it may hold "soft" (difficult-to-quantify) benefits before the public eye and even stimulate debate over their magnitude and how best to measure them. In any event, the second approach will be needed even for the "hard" benefits that are segregated out by the first approach because actual markets involve externalities and therefore their prices may need adjustment. The problem of externalities, it need hardly be said, is particularly acute in the environmental case. For example, Porter measures disposal costs for solid wastes simply by looking at what cities were paying per ton in the 1970s to have refuse hauled to the landfill and buried. A fuller accounting, however, would also consider what is now painfully evident: the limited space available for the purpose of solid waste disposal and the long-term costs of filling that space with nondegrading bottles and cans. Moreover, Porter's analysis looks at energy consumption in container production versus consumption in container recycling (because this measure is fairly readily gauged), but it ignores the resulting levels of pollution from production rather than recycling, the comparative environmental degradation in obtaining raw materials, and so on. These external factors are difficult to quantify and in any given case may be of little significance. The point is that they are easily overlooked even in very careful analyses, so that it is potentially misleading to summarize the results of such analyses as one-dimensional acceptance and rejection zones for policies. A legislator may believe he is looking at hard facts when he faces a $27 per year per citizen value of litter reduction necessary to make the deposit law a net benefit. In reality, he is looking at the easy facts—the ones most amenable to quantification. The reliability of an analysis should not be measured by how much its monetized values rely directly on market prices because the prices themselves may not reflect the true costs of a specific benefit.

A third approach may help remedy some of the problems of the first two: environmental policy might deliberately be used to reshape the boundaries of markets and to internalize external costs. This approach might involve a fairly elaborate governmental role in structuring some markets, but it would permit the use of market mechanisms to determine prices. Governmental intervention, then, is not always the enemy of a free market approach.

Within the realm of individual welfare, quantification of certain central benefits—for example, the aesthetic or social qualities of the environment within which one lives or works—may always elude analysts, even in cases in which market mechanisms have been modified to include externalities. Either the policy analyst must believe that these difficult-to-quantify contributors to welfare can be ignored—perhaps because they tend to cancel one another out, although this would be hard to believe in general—or he or she must be committed in advance to building many caveats into the final benefit-cost assessments and to bringing many more qualms and considerations directly to the attention of decision makers.

The five sources of possible inaccuracy that have been discussed—diminishing marginal utility in the intrapersonal case, diminishing marginal utility in the interpersonal case, preferences involving poor information or other cognitive defects, preferences not related to welfare, and the absence of appropriate markets—are all subject to some degree of correction, and this fact suggests something about how benefit-cost analysis might be improved as a measure of individual welfare. Of course, to take these difficulties seriously, it is necessary to take the notion of individual welfare seriously.

Economists sometimes air rather skeptical views about the concept of well-being, and one evident effect of such skepticism is to promote impatience with efforts to look closely at how adequate one measure of utility might be in comparison with another. I do not think that methodological or philosophical considerations require such skepticism, and as far as I can see the best case for defending the normative relevance of the enterprise of benefit-cost analysis depends in the end on rejecting it. It may be a deep-seated fact rather than a mere heuristic that microeconomics texts tend to begin with cardinal utility theory and to retain the vocabulary of utility even in the brave new world of ordinalism.

It would be possible, given the desire, for the government greatly to reduce the mystery about utility by greatly increasing the support given to experiments aimed at providing better information about well-being, as perceived through the lens of choice. Moreover, it would be possible, given the desire, for the government greatly to increase the support given to basic social and psychological research and to research into the toxicity or carcinogenicity of various chemicals, the behavior of substances in the atmosphere or the soil, the medical consequences of social dislocation, and so on. The United States justifies a vast military budget by pointing to the dangerous possibility of being caught unprepared. Yet even as I write, this nation is in danger of being caught unprepared in the realm of environmental policy making. At bottom the problem may not be the difficulty of doing good research on relating utility to willingness to pay but rather an unwillingness to pay for research with considerable utility. No doubt, this unwillingness is another example of a cognitive defect.

MEASURING COSTS AND BENEFITS OVER TIME

What does one have to believe to discount future costs and benefits? One concern that must be addressed if preferences are to be corrected for informational or cognitive defects is the fear that such correction will in fact result in a particular portion of the total population—in this case, technocrats—substituting their preferences for those of the population at large. Yet the very same sort of concern arises from certain uses of intertemporal discounting. For example, the use of intertemporal discounting is sometimes justified by pointing to a "social time preference"—people, it is said, prefer to have a given benefit occur sooner rather than later. (This phenomenon is not entirely uncontroversial, either as a descriptive[19] or as a normative matter, but let us assume it for now.) People alive today, then, will prefer benefits in the present. By symmetry, people of the future will prefer benefits in the future, which is their present. Thus, if existing time preferences are used to discount the benefits and costs of a social

---

[19] For example, people continue to set aside savings even when, as has often been the case in recent decades, the real after-tax rate of interest on savings is indistinguishable from zero (Lind, 1982:84). However, since people save for various reasons, one should be somewhat skeptical of using saving behavior too directly as a measure of time preference.

policy with long-term consequences, benefit-cost analysis will, in effect, be assessing the overall value of a policy by substituting the preferences of a particular portion of the affected population—those who are alive in the present—for the preferences of the population at large.

The above is not in itself a complaint of intertemporal injustice, although it could easily become the basis for such a complaint. Instead, the complaint is, in the first instance, one of mismeasurement, of failing to take into account the preferences of many of those on whom the consequences of the policy will fall. It is as if the costs and benefits of a national ambient air standard could be assessed only by consulting the preferences of those in Kansas. Indeed, the mismeasurement is worse than that because the time preferences of those alive in the present are not—unlike, say, those living in Kansas—even approximately representative of the time preferences of all those affected by the policy. People tend to show a fairly strong preference that benefits be received within *their* lifetimes, with costs postponed until sometime in the future. The comparison therefore should be with measuring the value of a national standard on acid rain by consulting the preferences only of those in the Ohio Valley.

Still, it might be argued that this comparison is misleading. If persons who are alive today are saving at an optimum rate, as they should be in an ideal market, then the welfare of future generations will be assured through economic growth. This situation is quite different from one in which people in the Ohio Valley benefit from cheap power while inflicting net losses on those downwind of them. Thus, the interests of future generations are not really being left out of account if people act on existing time preferences.

I would like to draw attention to two features of this rationale for following current time preferences.

When Saving Is Optimal

First, this rationale does not involve the discounting of future utilities. Under conditions of optimal saving, if people who are alive today were to ignore their own time preferences and force themselves to invest more and consume less, the total amount of consumption that everyone will enjoy over time would be reduced. This efficiency-based argument is temporally neutral; it functions to show that there is no real inconsistency in practice between, on the one hand, making decisions about consumption versus investment based on social time preference, and, on the other, counting all welfare effects equally, whenever they occur.

What role would discounting play in such a scheme?[20] Because future generations are assumed to be more prosperous than those alive today,

---

[20] I am indebted here to Allan Gibbard's comments on the original version of this paper.

they will be willing to pay more (in constant dollars) to receive a given benefit or avoid a given harm. Yet these higher, time-indexed, willingness-to-pay indicators will not reflect comparably large gains in utility, owing to the diminishing marginal utility of money. Discounting must be used to telescope these higher figures in a way that permits comparability with present willingness-to-pay indicators. This approach avoids the overestimation of future costs and benefits, which could result in mismeasurement of the relative value of competing policies that differ in terms of when their costs and benefits will obtain. From the standpoint of the present, a future monetized measure can be discounted to solve a problem of intertemporal comparison and obtain an accurate portrait of effects on utility over time. Once again, it is apparent that no discounting of future utilities—as opposed to prices—is involved.

Such discounting of monetized measures for the purpose of intertemporal comparison should presumably take as its rate not the real rate of economic growth but rather the rate at which the marginal utility of money is declining as such growth occurs. This rate can be estimated only roughly and, owing to disparities of wealth, will represent only an average within a population. Presumably, however, the upper bound is the real rate of economic growth because even at high finite levels of wealth, additional dollars will continue to have some positive utility. If growth is taking place and wealth differentials are not increasing, the lower bound will be zero. Thus, when discounting is used for the sake of intertemporal comparison, the usual rates employed in benefit-cost analysis, which often exceed the real rate of growth of the economy, are much too high and would be expected to yield systematic undermeasurement of future gains and losses.

Discounting for purposes of intertemporal comparison only makes sense if one is operating with time-indexed willingness-to-pay indicators. A similar effect could be achieved by using present willingness-to-pay indicators for future costs and benefits and then not discounting.[21] Disaster occurs, however, when one mixes these two strategies and uses present willingness-to-pay indicators for future costs and benefits and then also applies discounting. This approach produces the astonishing result that, after a few years, the outcomes of policies, no matter how harmful or beneficial, have almost no measured welfare effect. Such a result is hardly credible, for it would imply that it will not matter to people in the not-too-distant future how life goes for them.

It is possible that legislators or administrators who ask for benefit-cost analysis want nothing more than a measure of effects on the welfare

---

[21] The result may not be quite the same, however, because it is possible that, on average, members of future generations, owing to their higher standard of living, will receive more utility than members of generations alive today from, for example, a year of life.

of people alive today. It is also possible that, if benefit-cost analysis is performed in a climate of optimal growth, no real underrepresentation of future utility will result. Still, the function of benefit-cost analysis, as understood here, is to give the fullest possible accounting of welfare effects. Such an accounting might simply confirm the view that, in practice, acting on existing time preferences will not lead to policy choices that would in effect underrepresent future utility. But one would want to see this confirmation actually carried out.

This point brings me to the second feature of the rationale for following existing-time preferences that I would like to discuss.

When Saving Is Not Optimal

Currently, there is a vigorous debate about optimal saving, and I am in no position to judge it. Yet, it does seem difficult to find people who believe that, once all externalities have been taken into account, present rates of saving are optimal. Indeed, reflecting on the environmental problems of toxic wastes, atmospheric pollution and warming, nonrenewable resources, ozone layer depletion, and so forth, raises the real possibility that future generations will be worse off than today's. It is simply not clear that the economic growth made possible by rapid exploitation of the environment will be great enough to enable future societies to make up the losses connected with resource exhaustion and environmental degradation. To effect widespread ecological damage is quickly and rather cheaply done, but even the richest of societies do not appear to be able to afford to repair this damage fully without severe economic burdens.

One might (were it not an understatement of the difficulty) invoke what I am tempted to call the law of styrofoam pellets, a special case of the second law of thermodynamics: on a windy day it is the work of a moment to send styrofoam lentils all over hill and dale—I need only open a packing box and wave it aloft. To reassemble even nine tenths of the pellets in the original box, however, would give someone a week's work.

The wholesome and optimal expansion of the economy cannot be taken for granted. Without optimal saving and growth, the realm of operation becomes that of the second best, where one can no longer be certain that the effort to approximate an efficient solution in one area brings overall efficiency closer. So, most bets are off. In particular, because it is no longer safe to assume that people will be moving forever upward along curves of increasing utility, decision makers cannot ignore the problem of trying to gauge future utilities accurately in order to make comparative assessments of policies. To follow existing time preferences and employ a blanket discount function in the world of the second best is thus to adopt an attitude that carries a considerable possibility of underrepresenting actual welfare effects.

Competing environmental policies may differ greatly in regard to when they impose costs and how they affect the relative scarcity of certain goods. Although for practical purposes it may be necessary to limit how far into the future costs and benefits are traced, some remote consequences of present activity can even now be foreseen.[22] I recall as a teenager being disturbed in a way I could not quite express by a photo in the middle of an encyclopedia article on radioactivity. It showed a sailor pushing concrete blocks off the stern of a ship, and each block was full of highly radioactive waste. "The concrete blocks are expected to last 100 years," said the caption cheerily. Elsewhere in the same article, however, was the information that many of the radioactive isotopes present in such blocks have half-lives of thousands or millions of years. Policy makers can choose to make more or less sturdy concrete blocks, to make more or less expensive storage facilities, or to alter behavior so as to create more or less of the waste for disposal. The alternatives presented in these choices may mean cheap waste disposal in the present but an expensive or even irreversible problem in the future, or expensive waste disposal in the present and a cheaper and more reversible problem in the future. Benefit-cost analysis can help in determining the long-run expected value of such choices—but not by engaging in blanket discounting in a second-best world.

To be sure, discounting will be of use even so. One must, for example, be sensitive to the opportunity costs of capital, to worries about displacing investment, and to the compounding effects of taking benefits earlier rather than later. Yet in all such cases, future costs and benefits must be measured without actually discounting future utilities. Certain goods tend by natural (or "natural") mechanisms to multiply over time—rabbits and capital, for example. Consequently, if one wants to know the quantity of such goods that are needed on hand now in order to have some particular quantity of them on hand at a given time in the future, something like a present-value calculation based on the rate of increase can be carried out. This calculation, however, will not reveal how much of this good should be on hand at any particular future time, given present resources—that is a question one must try to answer by asking what will be needed or wanted, and how badly, at that time. This latter question supposes an evaluative stance of temporal neutrality, reflecting commitment to a full

---

[22] Actually, such limits may be necessary for theoretical purposes as well. In general, an infinite time horizon makes comparisons of costs and benefits impossible—both will simply grow indefinitely. Still, a distant but finite time horizon, along with some further constraints about foreseeable effects beyond that horizon (e.g., concerning something like the net rate of utility growth) would do a much better job of representing future utilities than straight discounting at any significant rate.

accounting of welfare effects. Such neutrality is consistent with discounting in regard to decisions about how best to deploy the "naturally increasing" instrumentalities of welfare.

This mention of natural increase through productivity gains brings the discussion to another wrong turn that may be taken in calculating opportunity costs. Practitioners of benefit-cost analysis have rightly rejected the idea that the value of a life is measured by its contribution to economic production and instead have developed measures based on willingness to pay. Such measures yield values that are seemingly more in line with intuitive views about the relative importance of life. Nevertheless, even though a particular sum—for example, $2 million—is set as the monetized value represented by a life, this value cannot be treated as if it were a $2 million investment. Suppose decision makers are asked to choose between two policies, one of which emphasizes saving lives in the present while the other emphasizes achieving some other benefit 10 years in the future. If the monetized value of a life were a monetary asset, it should be possible to calculate the opportunity cost of losing a life as opposed to saving it by considering the *return* on $2 million. This kind of approach however, would be misleading about the effects of saving a life in comparison with other benefits. Monetization is a potentially useful way of valuing various benefits for the purpose of comparison; it is not, however, a device for turning all benefits into interest-bearing assets that carry the present value of their monetized measure. It would be unfortunate if the hard-headed practicality behind monetization were to yield to mystification just at the point when its proper goal has been accomplished.

In sum, in this second-best world, decision makers cannot believe that they are entitled to carry out blanket discounting of future costs and benefits on the basis of an assumption that the welfare of future generations is being adequately protected by the underlying productivity of the economy. Policy assessment must endeavor to capture the actual expected long-term costs and benefits of today's choices as accurately as possible, and to weight them, using the best available estimates of the probabilities of outcomes rather than some uniform temporal discount. If measurement is to be consistent and accurate, it must use the same criteria for people alive in the future and their well-being that are used for those alive today. If consumers are sovereign, then they must be sovereign, whenever they come into being. One of the rationales for benefit-cost analysis is market failure; without optimal savings and growth, there can be no clearer example of market failure than the failure of the preferences of future generations to enter into the determination of prices today.

By providing a means by which to represent future preferences in present calculations, benefit-cost analysis can allow decision makers to confront the actual effects their choice will have upon welfare in the future.

Thus, the use of benefit-cost analysis not only is a matter of accurate measurement but could also involve the promotion of intergenerational justice by giving voice to temporally absent preferences. Benefit-cost analysts committed to a full accounting could become, in effect, representatives of otherwise disenfranchised future generations. It remains to be seen whether practitioners of benefit-cost analysis will seize on this—perhaps unexpected—opportunity to plant their analytic banner on the moral high ground.

ACKNOWLEDGMENTS

I would like to thank Allan Gibbard, Edward Green, Roger Noll, Robert Solow, and Hal Varian for helpful criticisms of an earlier draft of this paper. I am especially grateful to Hal Varian for suggesting useful readings and for helping me to see how to frame some of the issues I wish to raise in terms that might be intelligible to economists. If I have failed in this task, it is surely not his fault. I am also indebted to Douglas MacLean for numerous helpful discussions and for bringing to my attention Page (1977) and appendix F of Parfit (1984), both of which I found most useful in writing this paper.

REFERENCES

Ben-David, Shaul, Allen V. Kneese, and William D. Schulze
    1979   *A Study of the Ethical Foundations of Cost-Benefit Analysis Techniques.* Working paper. Washington, D.C.: Resources for the Future.
Conservation Foundation
    1987   Risk Management Case Study. Unpublished manuscript. Washington, D.C.
Leonard, Herman B., and Richard J. Zeckhauser
    1986   Cost-benefit analysis applied to risks: Its philosophy and legitimacy. Pp. 31-48 in Douglas MacLean, ed., *Values at Risk.* Totowa, N.J.: Rowman and Allanheld.
Lind, Robert C.
    1982   Introduction. Pp. 1-94 in Robert C. Lind et al., *Discounting for Time and Risk in Energy Policy.* Washington, D.C.: Resources for the Future.
Nozick, Robert
    1974   *Anarchy, State, and Utopia.* New York: Basic Books.
Page, Talbot
    1977   *Conservation and Economic Efficiency.* Baltimore, Md.: Johns Hopkins University Press.
Parfit, Derek
    1984   *Reasons and Persons.* Oxford: Clarendon Press.
Porter, Richard C.
    1978   A social cost-benefit analysis of mandatory deposits on beverage containers. *Journal of Environmental Economics and Management* 5:351-375.
Railton, Peter
    1986a   Facts and values. *Philosophical Topics* 14:5-31.

1986b   Locke, stock, and peril: Natural property rights, pollution, and risk. Pp. 89-123 in Mary Gibson, ed., *To Breathe Freely: Risk, Consent, and Air*. Totowa, N.J.: Rowman and Allanheld.

Rawls, John
1971   *A Theory of Justice*. Cambridge, Mass.: Harvard University Press.

Sen, Amartya
1985   Well-being, agency, and freedom: The Dewey Lectures 1984. *The Journal of Philosophy* 82:169-221.

5

Comparing Values in Environmental Policies: Moral Issues and Moral Arguments

Three kinds of problems seem particularly pervasive in administering environmental policy. The first is political. Policy makers work within a framework of environmental laws that are notoriously vague, apparently contradictory, and often otherwise flawed. Those charged with the responsibility to protect the environment must follow procedures that were designed in part to protect political interests; they must fight off relentless pressure while remaining accountable to politicians; and they must enact policies with an eye to withstanding the litigation that will surely follow. This process may be a model of democracy at work, but it does not encourage thoughtful responses to complex issues or regulatory improvement by "fine tuning" in the light of new information and greater experience.

The second problem in administering environmental policy is technical. Uncertainty is pervasive in this area, and the demand for precision is often greater than science or practicality permits. Under pressure to act now, decision makers grope for policies in ignorance, especially if the risks involved in various alternatives have latency periods between exposure and the onset of irreversible effects. The costs of always acting conservatively in the face of uncertainty can be prohibitively high; and it is not unusual for the uncertainties to be so great—ranging over six or seven orders or magnitude—that the risk analyses are useless to policy makers.

The third problem is one of comparability. It is impossible to eliminate environmental risks; policy makers must thus decide when to put more resources into reducing them, when to control other risks instead, and when to stop trying to reduce risks altogether and use resources for other

Douglas E. MacLean is professor of philosophy at the University of Maryland.

purposes. To make these decisions, a broad range of values must be weighed and compared. What kind of a problem is this? Insofar as it concerns the allocation of scarce resources, it is a classic economics problem, a general problem in rational choice or decision theory. Why should it be singled out as an especially central problem for environmental policies? It is surely not as immediate a concern for the policy maker as the political and technical problems he or she faces. The reason to single out this problem for environmental policies has to do with the nature of the benefits or values that are involved.

Environmental decisions typically require the comparison of different benefits (e.g., the preservation of human life, health, clean air and water, wilderness, endangered species, money, and consumer products). The economics problem might be described as finding the correct weights or rates of exchange among these benefits. That our society considers each of them valuable is obvious, but it is far from clear that there is a satisfactory metric for weighing them together and trading them off. All of the general methods proposed so far seem notoriously controversial. The underlying issue is the nature and value of different benefits, which is not strictly an economic issue but rather a philosophical one. It is a subject of ethics.

Some of the benefits that must be weighed and compared are even more difficult to evaluate than those already mentioned. For example, it is taken for granted that public policies must be fair and must consider the distribution of benefits, risks, and costs across locations or populations and across time. An important goal of policies is to protect the health and environment of future generations. Yet risks can be distributed differently in various ways, all of which may seem morally relevant. Some statistically certain number of annual deaths spread across a population may have to be compared with a small risk of catastrophic consequence. The average individual risks involved may be identical, but they seem to involve different values. How should they be compared? And how should these different distributions be compared with the other benefits and costs involved? These, too, are moral questions.

The philosophical issues involved in promulgating environmental policies, however, extend even beyond this. Many people believe it is simply wrong even to attempt to solve these measurement problems and to find a common and applicable rate of exchange between all of these benefits. One popular and persistent view holds that it is morally wrong to set forth the problem in this way. It is frequently alleged that the value of such benefits as human life or the nation's environmental heritage cannot be equated with money, or other resources, at any rate of exchange. To assign this kind of exchange value to such benefits is to treat them as commodities when they really have a different kind of value—a sacred value perhaps—and should be regarded as such. This objection suggests that it is necessary not only to find a method for weighing different goods but also to find appropriate ways of expressing or regarding different values. Part of the problem, it seems then, is procedural.

The philosophical issues at the core of this third problem are therefore quite basic. They involve not only the question of how to weigh different benefits but also the more general question of how to compare different values. The objections suggest that the benefits or values may be incommensurable, and yet decisions based on comparisons must be made.

This paper will address these moral issues, which are at the heart of many environmental controversies. Moral views drive debates over proposed legislation and are further reflected in subsequent litigation. The appropriateness of discounting the value of future lives, the application of benefit-cost analysis as a method of setting environmental policies more generally, and other issues that remain central and contentious in the environmental policy arena are essentially moral disputes.

METHODS OF REASONING ABOUT MORALITY

Most policy analysts would agree that the problems I have been describing are pervasive and at the core of many environmental disputes. My aim so far has been simply to suggest that many of these moral concerns raise a common general question about whether and how different kinds of benefits and values may be compared.

I wish to consider some of these moral issues directly, but it would be useful first to discuss briefly the nature of moral or philosophical reasoning. The approach I will take to these issues—the approach I find most natural and useful—is to inquire directly into the nature and implications of some of the values involved. What *is* the value of human life? What *are* current society's obligations to future generations? This process is not an empirical inquiry into what most people happen to believe but rather a normative inquiry. The question is, what is it reasonable to believe? A successful argument must appeal to what can be found to be reasonable and not merely what is persuasive. If different and incompatible views pass the test of reasonableness, then the moral inquiry turns to more general principles and procedures for making balanced or fair choices among these different individual views.

This kind of inquiry inevitably appeals to moral intuitions and to arguments that try to link such intuitions to other intuitions and to more general principles. Moral reasoning, as I would characterize it, is in this respect similar to scientific reasoning. Instead of reasoning back and forth between hypotheses and laws, on the one hand, and observations and experiments on the other, in moral reasoning, one reasons between cases and considered intuitions on one hand and more general concepts and beliefs on the other. Both kinds of reasoning are objective to the extent that their claims are susceptible to rational assessment.

One difference between scientific and moral reasoning, of course, is that moral principles are not meant to predict but to prescribe; but this difference is not crucial to the nature of the reasoning involved. If the nature of reasoning in science and ethics is similar, however, why is it so much harder to reach consensus in ethics than to reach consensus in science? One reason is that methods of reasoning are more highly developed and agreed upon in the sciences than they are in the field of ethics. Scientists can also design experiments explicitly to test hypotheses and resolve disputes. In ethics, one uses actual and hypothetical situations to test one's intuitions, but these "thought experiments" are much less likely to be convincing than are the experiments of science.

Such an approach to moral inquiry is common within the discipline of philosophy, but it appears to be alien and objectionable to policy analysts who are trained in other disciplines. They seem to favor other approaches to resolving fundamental moral issues. One of these approaches is empirical. The way to resolve conflicts that arise among analysts, it is sometimes thought, is to find out what people actually believe and prefer. This process will reveal what the nation's policies ought to be, for in democratic societies, the people are sovereign. I call this the low road to moral inquiry.

Another approach appeals to moral theories or general normative doctrines to resolve disputes and explain more particular values. Philosophers, of course, are also interested in moral theories and normative doctrines, but their interests often have very little to do with shedding light on particular moral or policy questions. Nevertheless, it is not uncommon to find policy analysts suggesting that this is the approach that philosophers might use most effectively to shed light on the moral aspects of environmental issues. According to this way of thinking, what is needed is a relatively comprehensive survey of the application of moral philosophy and ethics to problems of valuing risks to life and health, not a restricted set of specific arguments using a specific mode of philosophical analysis:

> Whereas no one could fault a philosopher concluding with an argument in favor of one perspective (e.g., Kantian) over others (utilitarian, rights libertarian, contractarian, etc.), what seems essential is to lay out what the alternative approaches are, what specific ethical and moral issues they address, and where the approaches diverge, as well as how one might assess the relative applicability and relevance of each.[1]

---

[1] This quotation comes from comments on an earlier draft of this paper by Roger Noll. Both the suggestion about how philosophical issues can be most usefully addressed and the list of philosophical perspectives or theories are common in the policy analysis literature (see, e.g., Kneese et al., 1983, and Keeney, 1984).

Let us call this the high road to moral inquiry.

The kind of direct moral inquiry I favor—using a restricted set of specific arguments—is neither merely empirical nor overly general or theoretical. It is a middle road. Let us briefly consider, then, the weaknesses in the high- and low-road approaches.[2]

Empirical Approaches to Moral Issues

Empirical approaches to these questions are usually attempts to uncover individual preferences. What do most people think are current society's obligations to future generations? How would most people trade off longer life expectancies against improved health, health against welfare, and so on? Revealed preference theory attempts to uncover these attitudes by looking at actual behavior and, in particular, at individual consumer choices in cases in which people are free and informed. The problem with this particular approach, of course, is that people are not always free or informed; even when they are, their consumer behavior does not always indicate their considered preferences, let alone their reflective moral judgments.

An alternative empirical approach is expressed preference theory or contingent valuation methods. These methods generally involve surveys in which people's preferences, their willingness to pay, and so on can be measured directly. Methodological difficulties aside (e.g., the reliability of surveys), a number of problems arise in applying the results of such surveys to settle moral disputes. The first is that people are normally asked in these studies what they would prefer, which is not at all the same as inquiring empirically into their moral beliefs. It is both consistent and common for people to have certain preferences that it would be wrong or unfair for policies to satisfy directly. Setting policies that attempt to satisfy the maximization of preferences would ensure, for example, that toxic wastes and hazardous technologies are always sited in the least populated areas simply because there are fewer people to object. Such a policy would obviously be unfair, and most people would agree that it is unfair, despite their personal preferences.

Recently, however, a number of contingent valuation studies have surveyed people's moral attitudes directly—in particular, their intuitions about procedural fairness and distributive justice (Kahneman et al., 1986). The information from these studies is interesting and perhaps quite useful, but there is a certain absurdity involved in thinking that these data can settle moral disputes. This absurdity is easy to show. Either moral truths can be reduced to individual preferences or they cannot. If they can be

---

[2] For a detailed discussion of different approaches to moral inquiry, see Parfit (1984).

determine what *is* fair or right—is the wrong question. Either people will answer according to which alternative they think is morally justified or best supported by moral reasons (which by hypothesis rests on an incorrect view of morality) or they will respond by saying what they think other people believe is right (if they have the "correct" moral views). Actually, to avoid relying on the false moral view one step removed, they would have to respond by saying what they think other people think that other people think that—and so on. If moral truths can be reduced to individual preferences, then soliciting opinions must either be ultimately self-defeating (if people express what they believe is morally justified) or land in an infinite regress.[3] If moral truths cannot be reduced to individual preferences, however, there is no apparent reason for surveying opinions to advance the understanding of moral truth and justification, except perhaps to help uncover reasons or arguments that might not be immediately apparent.

The foregoing is an argument against the low road to moral inquiry, but there are other reasons to respect popular sovereignty and in so doing uncover public opinion for use as a basis for policy decisions. One reason, of course, is that some comparisons may not essentially involve moral reasoning or moral problems. One might believe that moral philosophy really has little to say about how improvements in welfare should be traded off against improvements in health or life expectancy at the margin, when either improvement will fall to the same people (see Schelling, 1984). These choices may simply be matters of personal preference, but they are only some of the more troubling choices environmental decision makers must face. Many of the comparisons they make also involve distributional issues: improvements to one population must be balanced against risks to another, the price must be paid now to protect future generations, and so on. These, at least, are certainly moral issues.

Although some issues are simply matters of preference, there is a further argument for citizen or consumer sovereignty, based on the character of democracy, that applies to public policies. Marglin writes (1963:97), "Whatever else democratic theory may or may not imply, I consider it axiomatic that a democratic government reflects only the preferences of the individuals who are presently members of the body politic." The function of government, however, is not simply to reflect current preferences. Government also has an ennobling and educational role to play, even in a democracy. The U.S. Constitution protects many values, even those that are socially unpopular.

More specifically, it has frequently been suggested that government ought to have greater concern for the welfare of future generations than is expressed by individuals in their own choices (Pigou, 1932). Individuals

---

[3] I am indebted to John Broome for discussion of these points.

is expressed by individuals in their own choices (Pigou, 1932). Individuals die, after all, but the society continues, and there can be no objection to the government looking after the interests of future generations. This goal is explicitly stated in the National Environmental Protection Act and so should be a central concern of the Environmental Protection Agency (EPA).

For these reasons, it is proper for environmental policy makers to take moral arguments seriously, even when those arguments run counter to popular opinions. Some might object to this degree of "moralism" on the grounds that it is paternalistic. The argument against taking the low road, however, is moralistic only to the extent that it insists that moral conflicts be resolved through moral reasoning and not by empirical means. This approach is not necessarily paternalistic. Paternalism means restricting a person's freedom of choice or overriding personal preferences for the individual's own good. The argument against the low road does not call for replacing an individual's preferences for his or her own welfare with the decisions of "moral experts." It says only that moral decisions should be made that are justified by moral arguments.

Arguing From Theory or Basic Doctrines

Let us consider now the high-road approach to moral inquiry as it relates to environmental policies. This approach looks to moral theory or basic normative doctrines and attempts to apply them to particular policy questions.

Let us first consider libertarianism, contractarianism, and utilitarianism—which (as I indicated above) are sometimes thought to be different representative moral perspectives—in order to illustrate the difference between moral theories and basic normative doctrines. Libertarianism (or "rights libertarianism"), for example, usually appears in political theories that claim that individual rights are fundamental to determining political or moral obligations or claims. Even in the more comprehensive discussions of libertarianism (e.g., Nozick, 1974), libertarianism is at most a normative doctrine that is important in discussions of political theory.

Libertarianism is not a moral theory, because it does not take the kind of external view of normative claims that moral theories inevitably take. Discussions of libertarianism typically do not attempt to explain the foundation of its basic moral principles or relate them to other basic moral issues—for example, the nature of the subject matter of morality, the nature of moral reasoning, or the connection between moral principles and rational motivation or will. Moral theories address these more abstract issues, and they frequently are not very closely related to more particular normative claims or doctrines, unlike virtually all discussions of libertarianism.[4]

Contractarianism, in contrast, is a moral theory, at least as it has recently been developed (e.g., Rawls, 1971; Scanlon, 1982). Rawls's theory of justice as fairness, which is perhaps the best recent example of moral theory of any kind, does address these more fundamental and abstract issues. Although Rawls derives principles of justice (e.g., the principle of the priority of liberty, or the difference principle for justifying inequalities in the distribution of primary goods) that can be interpreted as normative doctrines, these principles are themselves intended to be applied only at a basic and rather abstract level. Rawls emphasizes that his principles of justice cannot be applied to immediate cases or particular policies.

It is important to emphasize this difference between moral theories and normative doctrines, because most moral theories appear to be compatible with many different moral principles or normative doctrines. A contractarian moral theorist, for example, might argue that a utilitarian principle is the normative doctrine at which a properly defined social contract would arrive.

In this respect, utilitarianism is relatively more complicated or confusing, because it has been defended both as a moral theory and as a normative doctrine or basic normative principle. As a moral theory, utilitarianism constitutes a particular view about the nature of moral goodness and moral justification or moral reasoning. Such a moral theory, however, may or may not prescribe the principle of utility as a normative principle. J.S. Mill, for example, did not regard utilitarianism as a moral principle that should be commonly adopted and applied directly to individual actions or to social policies. More recently, rule utilitarian and motive utilitarian theories have been suggested that defend other, commonsense moral principles and doctrines. Some of these theories even suggest that utilitarianism can be correct only as a moral theory, and that attempts to apply the principle of utility directly are likely to be self-defeating, from a utilitarian perspective (Scheffler, 1982).

Thus, if the high road means beginning with moral theory, it will require a long and philosophical journey before arriving—if at all—at recommendations for important policies and decisions. The high road would also require the resolution of disputes about which moral theory is correct. If these disputes are resolved, it will be in part only because one theory better explains all the moral phenomena, including more particular reflective moral judgments and the principles that unify them. In sum, it may be necessary to travel down what I am calling the middle road first in any case.

---

[4] For a good discussion of moral theories and moral doctrines, see Scanlon (1982).

If following the high road means instead beginning with basic normative doctrines—for example, some basic utilitarian principle, or libertarianism, or some egalitarian principle, or even the difference principle (although this may go beyond anything that might be attributed to Rawls)—then the question is, which normative doctrine should be chosen? A further task is to determine what in particular each normative doctrine means. Does the principle of utility call for maximizing pleasure, as it did for Bentham, maximizing some more abstract good, as it did for Mill and G.E. Moore, or maximizing the satisfaction of preferences, as it has come to be interpreted frequently among economists? If one chooses libertarianism, what do rights entail, which rights are basic, and how should conflicts among rights be resolved?

Surely, the only plausible way to answer these questions is to move to the more abstract realm of moral theory and to the realm of concrete reflective moral judgments about cases, policies, and principles. If all of the moral doctrines imaginable imply the same result in some case, then that consensus would be the strongest moral argument available, although it is very unlikely that there will be many cases like this, and, where they exist, they will probably not be in dispute anyway. Every moral doctrine, for example, must imply that gratuitous torture is bad; otherwise, people would immediately reject the doctrine. In general, if policy makers reason carefully and critically about more concrete cases and policies, they will have more confidence in their judgments about them than in their judgments about more general and sweeping principles or doctrines. Again, the middle road seems the place to begin and, for many practical purposes, the only road that need be traveled.

VALUING AND DISCOUNTING LIVES IN ENVIRONMENTAL POLICIES

Having defended the approach to moral reasoning that I find most natural and useful to apply to moral issues in environmental decision making, I will now illustrate this approach by considering some of the prominent questions I described at the outset of this paper.

Policy makers are usually interested in analyzing the values of the consequences of alternative policies. It is well known that some analytic methods that quantify costs, benefits, and risks and that aim at maximizing net benefits have difficulties evaluating different distributions of these effects. These methods take a narrow view of consequences.

Because a broader view of consequences would include the distributional effects of alternative choices, a sensitive metric should also, in principle, be able to evaluate distractions. The problem is that distributions may be valued differently because of their effects. Distributional principles cannot simply be valued independent of the context of their application. Rather, different principles and weights must be applied to different decision problems. Distributional values do not challenge analytic principles per se, but they challenge the generality of their application.

In addition to valuing the consequences of policies, certain procedures or ways of making decisions are also chosen. Elections are valued as a way of selecting public officials, trials as a way of determining criminal guilt, random processes as a way of achieving fairness. Some procedures are valued intrinsically and not only for the instrumental reason that they are most likely to produce the best outcomes. (There is generally little reason to believe they will.)

Some would argue that procedures, like distributional effects, should also be regarded as part of the consequences of environmental policies (Keeney, 1984). This argument is a far more controversial extension of the concept of consequences. I can illustrate why this is problematic with a personal example. Not long ago, I suggested to my wife that, instead of buying her a birthday present this year, I would give her money and she could shop for her own present. I am not very good at picking out presents for her, and I do not enjoy shopping for them. I offered her $50, explaining that I would probably spend $25 for a present if I bought it, so $25 could be considered to cover the cost of her time. Besides, she enjoys shopping for gifts. My wife did not appreciate this suggestion. She valued the traditional procedure for increasing her stock of goods on her birthday, and she did not regard money as compensation for this loss.

If one insists that the value of procedures is commensurable with other consequences, then it might be difficult to regard procedures as having more than instrumental value. In any case, an evaluation of a procedure must include a full assessment of the effects of using it, which is more than an evaluation of its expected consequences.

Two issues are central to moral debates over environmental policies. The first is the social discount rate, which raises distributional issues about how to compare costs and benefits that are spread out in time. The central moral issue here is whether future lives should be valued differently than present lives. This question leads to the second issue, which is how to compare the value of saving lives or improving health with other values and the suggestion that the best way to do this is to assign monetary equivalents to all of these values. This solution raises the problem of putting monetary values on human life and health. The moral objection to this approach is that the value of human life is incommensurable with other economic values: one cannot put a price on human life. I will argue that when some of the confusion is removed from this objection, the moral core of the issue involves procedural values.

THE SOCIAL DISCOUNT RATE

It is common in policy analysis to apply a discount rate to expected consequences as they occur further and further in the future. The reasons for discounting can appeal to the opportunity costs of capital, the reasons for wanting returns on investments sooner rather than later; or they can appeal to rates of time preference, the claim that people tend to care less about consumption in the future or about the remote effects of their actions and policies. These are very different kinds of justifications. Often, they are not distinguished, or they are not treated differently.

In a recent provocative study comparing the effectiveness of a large number of regulations, Morrall (1986) expresses a theoretically popular view about discounting. Explaining his own assessment techniques, he writes (p. 28):

> For the sake of consistency, I adjusted these temporal variations using a uniform 10-percent discount rate for both benefits and costs.
>
> Students of benefit-cost analysis will recognize an unavoidable imprecision in using a uniform discount rate, and a certain arbitrariness in using 10 percent rather than some other rate. Some regulatory costs displace investment and others displace consumption, and the two effects are not economically identical. Here as elsewhere, however, the analytical demands of tailoring a precise discount rate for each rule were impossibly large, and for comparative purposes the benefits of greater precision would have been small. Students of regulatory politics will recognize that discounting benefits as well as costs runs afoul of the policies of some regulatory agencies, not to mention the positions of some political representatives and op-ed writers. On this point my procedure is impeccable. Discounting costs but not benefits leads to absurd results, such as that a rule saving 100 lives a decade from now is more desirable than a rule of equal cost saving 99 lives right away, and that all rules yielding continuous benefits are worth any amount of immediate costs.

Morrall here suggests several different reasons why all benefits and costs should be discounted at the same rate.

Some of these issues arose recently with EPA's proposed asbestos rule because the detailed knowledge available about the health effects of asbestos allowed EPA to make relatively precise estimates of the latency periods between exposure and the onset of disease. Following its traditional practices, EPA proposed that it not discount the cancers averted for time, but the Office of Management and Budget (OMB) insisted on a 10-percent discount rate for all costs and benefits, which included discounting the latency periods between exposure to asbestos and the development of cancer. The issue is important for many other EPA concerns as well—for example, waste disposal regulations, an area in which much of the cost of regulating is borne initially but the possible costs of not regulating occur well into the future, and uranium mill tailing regulations, in which the yearly benefits of regulating are small but spread over a century or more.

Morrall does not explain why he chose 10 percent as the discount rate (10 percent is considered by many to be high), but in so choosing he follows recommendations urged by OMB since 1972 (OMB, 1972). The discount rate obviously has important political consequences. It has been the subject of considerable debate among practitioners and theorists who realize that the benefit-cost analyses of many development projects, as well as many regulatory proposals, are sensitive to small changes in the discount rate (Lind, 1982:1-18). I will try to disentangle some of the philosophical issues involved in discounting and treat them systematically.

SHOULD LIVES BE DISCOUNTED?

I have mentioned two justifications for discounting: opportunity costs and time preference. A third reason is uncertainty, which I will not consider here (see Parfit, 1983).

What justifies discounting lives saved or lost in the future? Clearly, lives do not have the properties of money: they cannot be invested or used to earn a rate of return. The idea that opportunity costs attach to the value of lives and not only to the resources invested in life-saving programs does not make sense.

The only justification for discounting lives would seem to be time preference or the discount rate on consumption. It must be this reason that leads Morrall to conclude that it is absurd to save 100 lives a decade from now rather than 99 lives right away. I wish to argue that it is not absurd to save more lives in the future than can be saved now, other things being equal, and that there are compelling moral reasons for doing so.

Suppose one knew that an action would produce two independent nonmonetary outcomes that have quite different values. The outcomes would occur several days or weeks apart, but it is not known which will occur first. Would that matter for an evaluation of the action? Except for considerations of anticipation and memory, which are additional, secondary effects, it would not. Mere differences in the timing of events do not matter.

The murder of an innocent 20-year-old woman is equally horrible if it happens today or tomorrow or 20 years from now. I take this to be nearly a self-evident truth. It is part of what it means to treat people as equals to regard these deaths as equally bad. I am not saying, however, that the murder of an innocent 20-year-old today or the murder of the same woman when she is 40 are equally bad. The difference in what she would lose is morally important. Yet when one compares the loss of a 20-year-old today to the loss of a similar 20-year-old in the future, the two losses are equivalent.

Other things being equal, therefore, if it is better to save 100 lives today than to save 99 lives today, then it is also better to save 100 lives in the future than to save 99 lives today. As a matter of psychological fact, many people might have a greater concern for the lives saved sooner, but this psychological fact is also morally insignificant. People suffer many prejudices, psychological biases, and weaknesses of imagination and will, but it is the point of moral reasoning to help them overcome these natural but unjustifiable inclinations. I believe Ramsey was right to dismiss these preferences as due merely to a lack of imagination (Ramsey, 1931). The situation might be different if there were other kinds of connections to the more proximate lives, for special relations can produce moral obligations. A person may be justified, for example, in devoting greater attention and resources to his or her own children than to strangers, but this position has nothing to do with time.

Policy theorists tend to be remarkably unpersuaded by this argument, except perhaps to regard the conclusion as a public attitude widely enough held that it must be taken into account. I will try, therefore, to respond to four objections to it.

Democracy and Consumer Sovereignty

Some have argued from the assumption of consumer sovereignty that government policies should include discount rates that reflect society's time preference. Earlier, I discussed the relevance of arguments about democracy to moral argument and so will add a few brief comments here.

Three different questions should be distinguished: (1) As an individual, am I morally justified in being less concerned about the welfare of people in the future than about the welfare of those alive today? (2) As a community, are people morally justified in being less concerned about the more remote effects of social policies than about the more immediate effects? (3) If most people are less concerned about the welfare of people in the future, ought the government to override this majority view?

Consumer sovereignty is relevant to the third question. The moral inquiry being conducted here focuses on the first two questions about the justification of people's values and normative beliefs (Parfit, 1983). The question, therefore, is whether consumer preferences or popular opinions constitute a promising starting point for moral argument or whether they should be taken by themselves to be a moral intuition with normative status.

The preferences and the more considered values of individuals may differ considerably. Individuals might show time preference in their consumer behavior and investment decisions but different values and a greater concern for the future in their political choices (Sagoff, 1986). Marglin (1963:97) considers this possibility to be schizophrenic and argues that, in any case, consumer behavior is a better indicator of a person's "real" preferences: "[S]ince deeds speak louder than words, one can argue that preferences revealed in the market place are more genuine and better considered."

I find it remarkable to dismiss political activity as merely words. I would guess that most people are more committed to their political beliefs as an expression of their values than to much of what they buy as consumers. Political activity takes many forms. It can consist of supporting environmental groups that lobby for preservation while consuming nonrenewable resources beyond one's strictest needs. It is not clear whether there is even an inconsistency involved in this pattern of behavior, but it is surely not schizophrenic. It might be true that the values people express in political contexts and the programs they support tend insufficiently to take economic considerations into account. People find it easy to support a worthy program when they do not know what it costs. Yet it is perhaps equally likely that people do not take their moral values sufficiently into account when they shop or invest. Even if there were inconsistencies between market-exhibited preferences and those that are politically revealed, it is far from obvious that they should all be resolved in one direction.

These issues demand a much fuller treatment than can be provided here. I am suggesting, however, that the connection between consumer sovereignty and democracy be viewed with some suspicion. Sovereignty in a democracy rests with the citizens—people—and people express their values in many ways.

Excessive Sacrifice

The remaining three objections do not challenge the moral argument directly but claim instead that certain absurdities follow if all benefits, including the value of human lives, are not discounted in benefit-cost analyses. Morrall points to one such absurdity when he suggests that "all rules yielding continuous benefits are worth any amount of immediate costs" (1986:28).

Situations in which this result might follow are familiar at EPA. For example, EPA's analysis of a uranium mill tailings standard estimated that the present costs would be $388 million and that the standard would save 4.9 lives per year perpetually. If lives were discounted at 10 percent, the present value would be $8 million per life saved, a figure that is probably too high to recommend. If lives are not discounted, the cost-effectiveness is $800,000 per life saved for a horizon of 100 years, $80,000 per life saved for a horizon of 1,000 years, and so on.

Even in this example, the result of not discounting lives is absurd only with the assumption of an infinite time horizon, and there are good reasons to avoid such an assumption. The uncertainties involved should probably lead to the rejection of even the analysis with the 1,000-year horizon; yet however a reasonable horizon is determined, there is no justification for discounting the values of the lives analysts believe can be saved.

If the suggestion of selecting a time horizon in such cases seems ad hoc, there is another way to avoid the absurdity, using a method that has a stronger conceptual foundation: limiting the sacrifice required of the present generation.

It is common to think that individuals have less responsibility to do good than to avoid doing harm because avoiding harm can often be accomplished by not doing anything, whereas doing good requires some effort. There are limits on the effort that can be demanded of people on behalf of morality. It would not be wrong for an individual to avoid saving even hundreds of lives if that act required the sacrifice of all other benefits in that person's own life. It would be extremely noble for a person to devote himself or herself to such a morally worthy cause, but society cannot require this great a sacrifice. Individual moral rights may be seen in part as an effort to protect individuals from the possibly excessive demands of morality. There are similar limits on the sacrifices that can be demanded of current generations for the benefit of future generations. Identifying these limits blocks the absurdity noted by Morrall. It should be no more difficult to identify a maximum level of acceptable sacrifice for a generation than it is to determine an optimal social rate of savings.

Indefinite Delay

Let us consider the following example.[5] Certain social resources are available and can be used for either of two projects and in no other way. The projects have the following potential results:

- Project A will reduce by 100 the number of fatal accidents occurring in the 10th year following its initiation.
- Project B will reduce by 500 the number of fatal accidents occurring in the 40th year following its initiation.

It follows from the conclusion that the value of future lives should not be discounted that Project B should be chosen. Yet the final two objections I wish to consider claim that, if several plausible assumptions are added to this example, other absurd consequences will follow.

---

[5] I owe this example to Robert Dorfman; see note 6 below.

Let us first assume that the resources do not have to be invested immediately and that the identical projects will remain available in the future. Suppose the resources can be invested at a 6 percent rate of return. If policy makers decided to save for 10 years and then invest in Program B, the number of fatalities in the 50th year would be reduced by 900. Let us call this Program C. If discounting is not used, decision makers should prefer Program C to Programs A or B because it saves more lives. Then, however, decision makers should actually prefer Program D (which says wait 20 years and then choose Program B) to Program C, and so on. By this line of reasoning, investment would be delayed indefinitely.

This objection makes one plausible assumption—that there are usually alternative possible uses for resources—but it also makes some implausible assumptions. The resources would have to be invested and not consumed; then, they would have to be reinvested, with interest, in the life-saving program in the 10th year. Will programs be available at the same cost in the future? Will policy makers have the resolve to use the resources that have been set aside for this purpose? The half-life of political commitments is considerably shorter than the half-life of uranium mill tailings. Furthermore, how will science and technology have changed in this period? It is unreasonable to assume that the situation will remain the same indefinitely. Policy makers have good pragmatic reasons to do what they can in the present and to choose programs they can implement immediately, but these are not reasons to save lives immediately. It is not uncommon at EPA for analysts to develop programs that can be implemented now but that will save lives mostly in the future.

A Paradox

There should be discounting for opportunity costs with money invested at some rate of return. Therefore, monetary costs should be discounted. In deciding, however, whether a life-saving program is cost-effective or cost-beneficial, it is necessary to assign a monetary value to the lives saved. (I will discuss the moral issues involved in this process later.) These assumptions appear to generate a paradox: the value of lives saved should not be discounted but money should be discounted; nevertheless, the lives saved can be given monetary equivalents.[6]

Let us refer again to Projects A and B and consider now an alternative, Project C', which is to choose neither program and return the resources to the private sector where they will be consumed or invested according to

---

[6] I first discussed this paradox at a workshop on discount rates at Resources for the Future in 1985. Robert Dorfman reformulated it more precisely and elegantly than I had. I rely here on his reformulation.

individual savings preferences. Suppose the present value of these resources is $100 million and the discount rate for investment and consumption is 6 percent.

Suppose further that it is determined that a life saved is worth $2 million. Program A would confer benefits worth $200 million in the 10th year; Program B would confer benefits worth $1 billion in the 40th year. The present value of Program A is $112 million, and the present value of Program B is $97 million. Without Program C', Program B is preferable to Program A. With Program C', Program A is preferable to Program B, and Program A passes a benefit-cost test while Program B does not. This result seems to violate the independence of irrelevant alternatives.

One of our assumptions must go: either that the value of lives should not be discounted or that the value of a life has a monetary equivalent. It is the latter assumption that should be given up.

There are no opportunity costs attached to saving lives later rather than sooner in the example. If these benefits are assigned monetary equivalents, they should not be discounted to a present value. One might then say that Program B confers undiscountable benefits worth $1 billion when they occur; thus, they are worth that much today, but even this formulation is very misleading. One should instead regard the alternatives as not investing $100 million worth of resources in life-saving programs, investing the same resources to save 100 lives, and investing them to save 500 lives. Program B would save lives as a cost of $200,000 per life—at today's rates, a bargain. Program B should be chosen.

Variations of these last three objections are common in the literature defending discounting for time preference in benefit-cost analysis. None of the authors in the various articles, however, gives a good reason for discounting the value of lives saved or lost in the future as a result of present policy choices and investments. Similar arguments would apply to the value of improved health and perhaps to other kinds of benefits as well. These objections show that benefit-cost analysis must be applied carefully; in addition, selectivity must be exercised in choosing what to discount, determining time horizons for the analysis, assigning monetary equivalents, and so on. I am arguing for selectivity and not for rejection of the method of analysis or discounting; that is, in cases in which discounting is appropriate.

WHAT SHOULD THE DISCOUNT RATE BE?

I have argued that some costs and benefits should not be discounted. Other costs and benefits should be discounted, but at what rate? With a 10 percent discount rate, the present value of $1,000 of benefits 50 years hence is $8.52; at a 5 percent rate, the present value in 50 years is $87.20; at a 2 percent rate, it is $371.53. For a 50-year horizon, the difference between 10 percent and 5 percent is an order of magnitude; the difference between 10 percent and 2 percent is a factor of 40. The discount rate chosen can easily determine in many cases whether a project is cost-beneficial. Not surprisingly, therefore, an enormous technical literature argues for different rates.

I will not try to assess this literature here. I will instead suggest only why it might be wise to choose a social discount rate for health and safety regulations that is lower than the private sector rate; that is, lower than the discount rate individuals and firms apply to investment decisions based on rates of return and rates of interest prevailing in the market.

The principal reason given for similar social and private discount rates is the belief that social investments should meet the same economic standards that private investments must meet. This belief is sometimes reinforced by a currently popular political theory, which says that the proper role of government is to enforce rights and to act to correct market failures. Many people would reject this political theory and defend a more positive role for government. Indeed, the laws that give EPA its mandates make little reference to correcting market failures or to treating health and the environment as economic resources.

Even those who do not accept the market failure concept of government, however, ought to favor policies that make some economic sense, and this rationale is a sufficient reason to apply discount rates to some aspects of public investments. In some cases (e.g., development projects), the policy goals are economic; in some development projects, government investments will be competing with private sector investments. In these cases, there are better reasons for applying the same criteria private investors apply so as not to displace private investment.

This line of reasoning applies most clearly to water and energy projects. It does not apply to environmental, health, and safety regulations, areas in which there are both little evidence that the problems are caused by market inefficiencies, and independent moral arguments and political support for government action. If one also accepts that government has a special responsibility to protect the interests of future generations, a responsibility individuals and firms do not have and do not reflect in their economic decisions, then the reasons for applying different criteria to public and private investments are even stronger. The argument that regulations should be justified in economic terms is weaker; consequently, so is the argument that the social and private discount rates should be the same. Private investors do not worry that some resources are nonrenewable and that some costs and benefits are not replaceable. They do not worry that Americans have moral objections to treating certain benefits—such as human lives or perhaps wilderness areas—as resources to be exploited and invested. Citizens do worry about such things, however, and that is why laws are passed to establish regulatory agencies. It is natural to expect these agencies to operate with standards that are different from those applied in the marketplace.

There is another, related argument for lower social discount rates that is much discussed in the literature. This is the isolation argument (Marglin, 1963; Sen, 1967, 1982), which claims that the social discount rate should be lower than the private rate, even it the social rate is based strictly on individual preferences and all individuals have the same private rate of time preference.

The isolation argument assumes that individual savings decisions are based in part on an individual's altruistic preferences for future generations, perhaps especially for his or her own descendants. Yet individuals have limited control over the benefits these later generations will receive because they cannot control the investment decisions of others. Thus, individuals might choose to save more, or to have a lower discount rate on consumption, if they could be assured that their greater savings would be matched by the greater savings of others. This case is a variation of a common public benefits problem, in which government policies for protecting the interests of future generations act as the coercive mechanism needed to secure the cooperation of others. Thus, a discount rate lower than that used by individuals in isolated decisions should be applied to national environmental policies.

As I have presented them here, these arguments for different private and social discount rates are obviously neither conclusive nor prescriptive (they do not say what the social discount rate should be). My point is only to suggest how moral arguments might be applied to determine not only which future costs and benefits should be discounted but also what discount rate should be applied.

PUTTING A PRICE ON LIFE

Regulatory agencies do not set a price on human life; they merely try to uncover public preferences for risk reduction from data about consumer safety decisions, wage rate differentials for hazardous occupations, surveys, and contingent valuation studies. This approach is somewhat misleading because another possible way to determine the value of life would be to examine regulatory decisions themselves. Yet the norm is sought in other areas, of course, precisely to guide these decisions and evaluate existing regulatory policies. Such guidance and evaluation are the sole purposes in determining a social value of human life. This issue is a source of enduring controversy. Pricing life seems necessary for both holding regulatory agencies economically accountable and at the same time finding them morally repugnant.

The moral issue, of course, is treating a "sacred" good as an economic commodity. Kant (1785:Ak. 434-435) wrote that human beings have intrinsic worth or dignity and that whatever has dignity is "above all price, and therefore admits of no equivalent." Solow, however, describes the issue more clearly: "It may well be socially destructive to admit the routine exchangeability of certain things. We would prefer to maintain that they are beyond price (although this sometimes means only that we would prefer not to know what this price really is)."

Decisions about acceptable risk inevitably involve comparing the value of saving lives and protecting health with the costs of doing so. Many would argue that, to make these decisions in a fully rational way, it is necessary to be explicit about these costs and the amount our nation is willing to spend to save lives. Such disclosure is to adopt a moral position about procedures for making important decisions, a position that Gibbard (1986:99) calls one of "technocratic moral reform." He suggests that

> a rationally grounded morality will be reformist—perhaps shockingly so. It will not appeal primarily to our capacities to be aroused, as traditional moral reform movements have done, but to ways of regimenting the considerations involved to produce rational, coherent judgments. . . . We need, it seems, to train people in rational methods of risk assessment and so organize society that those methods really do determine policy with regard to risk.

Why do "many, perhaps most of us," as Gibbard says, find this a chilling prospect?

The answer is not, as some have claimed, that human life has an infinite price. That idea is not what Kant meant in saying that humanity is "above all price." The reason is rather, as Solow observes, that civilized people find it morally repugnant to view life as routinely exchangeable for other benefits. People thus feel uncomfortable with even rationally defensible procedures for making difficult decisions if those procedures make finding an exchange rate for life a prominent feature.

What sense can be made of this reaction? I would argue that the value of life is complex.[7] Human life has intrinsic value, which makes it worth saving and prolonging. This component of life's value favors efficiency and the saving of more lives rather than fewer. It stands behind support for rational methods of risk assessment. Yet human life is also sacred, and this component of its value can work in a different and conflicting direction.

Durkheim (1915) regarded sacred values as "elementary forms" of religious life, by which he meant that even as societies become secular,

---

[7] I have argued this point more fully (MacLean, 1983, 1986).

there remains a kind of need that traditional religions fulfill in older or more primitive cultures. This need involves finding rituals that strengthen social integration. "There can be no society," Durkheim wrote (p. 417), "which does not feel the need of upholding and reaffirming at regular intervals the collective sentiments and the collective ideas which make its unity and its personality."

As a pluralistic culture, the United States does not have a single unity and personality, but there are clearly some basic moral values that all Americans share and that, as Durkheim would point out, are universal. Rituals carry symbolic meaning that call attention to these values. They are marked by special, perhaps nonrational behavior and actions that draw the attention of the community to objects or relationships that have a special place in the life of the group. Because rituals are symbolic, they rely on conventional forms of behavior, which can differ from group to group. In all societies, however, and especially in those areas in which man finds it necessary to "humanize" parts of his existence, characteristic activities marked by rituals can be found (Hampshire, 1983). These rituals surround birth, sex, and marriage, for example, and they also surround death and the taking of life.

Precisely because health and safety decisions have obvious economic consequences, it is necessary to guard against treating human life as exchangeable in these contexts. Some policies and procedures that are inefficient but highly symbolic can be effective guards. Startling examples of inefficient, ritualized behavior are common in our dealings with hazards and risks. For example, Americans are generally willing to engage in rescue missions when identified individuals are involved and to act as if—or certainly to give the appearance that—costs are not a consideration.

To argue that the sacred value of human life in these situations must be respected is not to deny in any way the value of efficiency in lifesaving and the importance of saving more lives rather than fewer whenever possible. The point is rather a more subtle one. It is to suggest that there may be irresolvable tensions between our rationalistic, revisionist sentiments, on the one hand, and our conservative, ritualistic sentiments on the other. A rationalistic decision procedure may unavoidably threaten some of these sentiments, which may suggest that it is better not to make that procedure too absolute, too open, or too openly identified with public agencies like EPA that were created to pursue moral as well as other goals. It may perhaps be necessary to live with some controversies rather than to resolve them technocratically, and to tolerate "pockets" or modest levels of inefficiency for this purpose.

I mean also to suggest that the symbolic role of public figures like the EPA administrator should be recognized when he or she appears at a press conference to announce a regulatory decision, often about some hazard that to be reassured that things they value deeply—health, the environment, posterity—are being guarded and protected by the agency that has been established as the trustee of these values. Like it or not, the actions of EPA have important symbolic and expressive significance. Everyone may not approve of such taboos as refusing even to look at benefit-cost analyses, but it is necessary at least to be sensitive to the kinds of symbolic importance they might have.

CONCLUSION

The issues I have discussed involve very different kinds of arguments that lead to different paths of moral inquiry. Nevertheless, I think they all support the general conclusion that it is important to look critically and perhaps even suspiciously at suggestions that some analytic method should be universally applied to environmental decision making. The comparisons and trade-offs that must be made are often context dependent or may involve symbolic elements. These comparisons make the justification of decisions specific to a particular situation.

It is extremely important to use analysis to organize complex data and make decisions more consistent and efficient. It is also important, however, to realize that different values may have to be treated differently. It is this fact, and not measurement problems, that makes value comparisons in environmental policy making so difficult.

ACKNOWLEDGMENT

Many people commented on earlier versions of this paper. Claudia Mills, Talbot Page, Amartya Sen, and Susan Wolf provided some particularly useful criticisms. A. Myrick Freeman, III, and Roger Noll sent me detailed written criticisms. I have benefited from, though I am aware that I have not fully responded to, the questions they raised.

REFERENCES

Durkheim, E.
    1915   *The Elementary Forms of Religious Life.* Translated by J.W. Swain. London: George Allen and Unwin.
Gibbard, A.
    1986   Risk and value. Pp. 94-112 in D. MacLean, ed., *Values at Risk.* Totowa, N.J.: Rowman and Allenheld.
Hampshire, S.
    1983   *Morality and Conflict.* Cambridge, Mass.: Harvard University Press.

Kahneman, D., J. Knetsch, and R. Thaler
1986   Perceptions of unfairness: Constraints on wealth seeking. *American Economic Review* 76:728-741.
Kant, I.
1785   *Grundlegung zur Metaphysik der Sitten.* Page references to Prussian Academy edition of Kant's works (1911). Vol. III. Berlin. Translated by J.W. Ellington as *Grounding for the Metaphysics of Morals* (1981). Indianapolis: Hackett.
Keeney, R.L.
1984   Ethics, decision analysis, and public risk. *Risk Analysis* 4:117-129.
Kneese, A.V., S. Ben-David, and W.D. Schulze
1983   Ethical basis of benefit-cost analysis. Pp. 59-76 in D. MacLean, ed., *Energy and the Future.* Totowa, N.J.: Rowman and Littlefield.
Lind, R.C., ed.
1982   *Discounting for Time and Risk in Energy Policy.* Washington, D.C.: Resources for the Future.
MacLean, D.
1983   Valuing human life. Pp. 89-107 in D. Zinberg, ed., *Uncertain Power.* New York: Pergamon.
1986   Social values and the distribution of risk. Pp. 75-93 in D. MacLean, ed., *Values at Risk.* Totowa, N.J.: Rowman and Allenheld.
Marglin, S.A.
1963   The social rate of discount and the optimal rate of investment. *Quarterly Journal of Economics* 77:95-111.
Morrall, J.F.
1986   A review of the record. *Regulation* 10(Nov/Dec):25-34.
Nozick, R.
1974   *Anarchy, State, and Utopia.* New York: Basic Books.
Parfit, D.
1983   Energy policy and the further future: The social discount rate. Pp. 31-37 in D. MacLean, ed., *Energy and the Future.* Totowa, N.J.: Rowman and Littlefield.
1984   *Reasons and Persons.* Oxford: Clarendon Press.
Pigou, A.C.
1932   *The Economics of Welfare.* 4th ed. London: Macmillan.
Ramsey, F.P.
1931   A mathematical note on saving. In F.P. Ramsey, *Foundations of Mathematics and Other Logical Essays.* London: Kegan Paul.
Rawls, J.
1971   *A Theory of Justice.* Cambridge, Mass.: Harvard University Press.
Sagoff, M.
1986   Values and preferences. *Ethics* 96:301-316.
Scanlon, T.M.
1982   Contractualism and utilitarianism. Pp. 103-128 in A. Sen and B. Williams, eds., *Utilitarianism and Beyond.* New York: Cambridge University Press.
Scheffler, S.
1982   *The Rejection of Consequentialism.* Oxford: Clarendon Press.
Schelling, T.C.
1984   The life you save may be your own. Pp. 113-146 in *Choice and Consequence.* Cambridge, Mass.: Harvard University Press.
Sen, A.K.
1967   Isolation, assurance and the social rate of discount. *Quarterly Journal of Economics* 81:112-124.

1982   Approaches for the choice of discount rates for social benefit-cost analysis. Pp. 325-353 in R.C. Lind, ed., *Discounting for Time and Risk in Energy Policy.* Washington, D.C.: Resources for the Future.

Solow, R.
1981   Defending cost-benefit analysis: Replies to Steven Kelman. *Regulation* 5(Jan/Feb): 39-41.

U.S. Office of Management and Budget
1972   *Circular No. A-94.* (Revised) To the heads of executive department and establishments, subject: discount rates to be used in evaluating time distributed costs and benefits. Washington, D.C.: OMB, March 27.

6
Environmental Policy Making:
Act Now or Wait for More Information?

JEFFREY E. HARRIS

Environmental policy making is a dynamic process. Rarely do regulatory agencies make once-and-for-all choices between action and inaction. Instead, they choose, again and again, between degrees of action and waiting; making decisions that are based on information—scientific, economic, political—that changes continually.

This dynamic quality of environmental decisions poses serious problems for benefit-cost analysis. To evaluate a contemplated regulatory intervention, it is no longer enough to compare the intervention's currently estimated benefits and costs. In fact, it is insufficient to assess the whole future stream of expected benefits and costs. Environmental decisions also require estimates of the benefits and costs of regulating in the future as opposed to acting now. If the regulatory agency decides to act now, its experience with implementation may be informative about the costs and benefits of later policy choices, including future rescission of the regulatory action. In deciding to act now, the environmental decision maker thus needs to assess the future benefits and costs of correcting or rescinding policy mistakes.

The idea that policy choices are dynamic is hardly new. Most public policy decisions—in fact, most individual decisions—are dynamic ones. When a public utility commission disapproves a requested rate increase, it contemplates the benefits and costs of approving the increase later on. When stockholders decide not to sell their holdings, they consider the benefits and costs of selling later. The same is true for individuals who are seeking another job or deciding to go on a diet.

Jeffrey E. Harris is associate professor, Department of Economics, Massachusetts Institute of Technology, and physician, Primary Care Program, Medical Services, Massachusetts General Hospital.

Environmental policy, however, is an extreme case of dynamic decision making because regulatory decisions about environmental hazards are routinely made in the face of huge uncertainties—uncertainties in estimates of health risks, mechanisms of disease, the extent of exposure, or the costs of risk control. Under such extreme uncertainty, the appearance of even a modicum of new data can swamp the decision maker's prior beliefs concerning the costs and benefits of regulatory intervention. As a result, regulatory action on suspected hazards can be triggered or stifled by the issuance of preliminary toxicological findings, by false alarms concerning the measurement of environmental contaminants, or by leaks of draft reports of blue-ribbon panels.

In the conventional research models, repeated measurements tend to improve the precision of estimates of benefits and costs. With the extreme uncertainties encountered in environmental decisions, however, new research findings can pose unexpected contradictions, thus enhancing rather than reducing uncertainty.

My task in this paper is to explore, at least in a preliminary way, these dynamic complications of environmental policy making. My method of analysis is essentially anecdotal; that is, I offer some generalizations and then cite selected case studies for support. The hypotheses put forward in this paper need independent and more systematic testing using a representative sample of decisions faced by regulatory agencies.

In the next section, I establish the central, paradigmatic problem in the dynamics of environmental decision making—that is, the problem of timing. Do we act now, or do we wait for more information? The frequently voiced preference for waiting, I would suggest, is based upon a strong but unstated assumption: environmental policies are irreversible, and interventions by regulatory agencies impose large, sunken costs on private firms and consumers that cannot later be recovered.

I then inquire further into the realism of the irreversibility assumption. I find that in many cases, a contemplated environmental policy can grow more irreversible with continued delays. There are two mechanisms for this phenomenon of growing irreversibility. First, an environmental problem in its early phases may be amenable to partially reversible interventions (e.g., restrictions on use or access, product labeling, or pollution fees). If the problem gets worse later on, however, then truly draconian, irreversible actions may be required. Second, regulation is a game between governmental agencies and the private sector. The longer the regulatory agency delays action, the more time private agents have to make large, sunken investments in the prevailing technology. If the agency delays too long, the stakes become too high.

In a subsequent section, I probe further into the issue of "research." Although a strategy of delay is often coupled with a decision to invest in new data collection, I suggest that research is just as compatible with regulatory intervention. In fact, some regulatory actions are themselves a form of research because they provide essential information about the benefits and costs of future regulatory decisions. In principle, regulatory action can often be a better investment in knowledge than pure research without intervention.

I thus propose that policy makers consider two types of questions when contemplating the benefits and costs of a proposed regulatory action: How irreversible is intervention? How informative is the intervention? In general, my analysis points toward a style of regulation in which agencies take small, incremental regulatory steps at the early stages of a problem. These small steps would be designed to impose minimal sunken investments in compliance and still provide essential information on the uncertain benefits and costs of intervention.

IRREVERSIBILITY AND THE BIAS TOWARD WAITING

All too often, one hears the following refrain from scientists and policy makers: "We do not yet have sufficient information to take regulatory action. We would prefer to wait for better data to come in. We need more research."

This bias in favor of waiting and against action has been articulated in many forms. The following examples are illustrative.

> It may be that a proportion of lung cancers in man are induced by tobacco smoke; at the moment we do not know, but let us be sure of our evidence before we scare our public. (Passey, 1953)

> Thus, I conclude that in my personal view, given the current information, the banning of saccharin at this point in time is counterproductive, and I believe the ban should not be instituted until or unless some "safer" nonnutrient sugar substitute is available. (Isselbacher, 1977)

> DES [diethylstibestrol] could have been taken off the market immediately, without a hearing, if the FDA [Food and Drug Administration] had declared it to be an imminent hazard to health. That is the only statutory basis for immediate withdrawal of a drug from the market without first offering a hearing. The agency went to the National Cancer Institute [NCI] on this issue, and the NCI said that, in its judgment, DES was not an imminent hazard. The government's own scientists concluded that the risk was not of that magnitude. Therefore, there was no legal basis for taking that action. (Hutt, 1977)

> EPA [Environmental Protection Agency] did not immediately suspend these uses [of ethylene dibromide as a grain and fruit fumigant] despite the carcinogenic potential because EPA management did not believe enough was known at the time about the risks from residues on food,

the risks from substitute fumigants, or the risks from leaving crops and
foodstuffs unprotected. . . . It decided to await the results of studies
then in progress. (Russell and Gruber, 1987)

Each of these statements is a variant on the same basic theme: immediate action may be too costly in comparison to waiting. In Passey's
view, the costs arose from scaring the public. For Isselbacher, the cost
would be the absence of an alternative to saccharin. In the case described
by Hutt, it was too costly to bypass standard regulatory procedure and
ban diethylstilbestrol without a hearing. Russell and Gruber's discussion of
ethylene dibromide suggested several types of costs, including the risks of
substitutes for ethylene dibromide (EDB).

All of the examples contain an implicit benefit-cost calculation. The
benefits of a determination that smoking causes lung cancer, Passey argued,
did not outweigh the costs of "scaring" the public. The cancer risks of
saccharin, Isselbacher contended, were outweighed by its benefits as a
nonnutritive sweetener.

There is more to each of these examples, however, than a one-time
benefit-cost analysis. In each case, the decision to act or wait recurred. In
analyzing the benefits and costs of action and inaction, each writer needed
to consider how such benefits and costs might change over time. The
benefits and costs of action were really the benefits and costs of acting
immediately as opposed to acting later.

Thus, Hutt's description does not imply that DES carried no danger but rather that, in NCI's opinion, the danger was insufficient to act
immediately. Isselbacher likewise did not deny saccharin's cancer-causing
potential. Instead, he urged action later, once a substitute was available.
EPA did not deny the carcinogenicity of EDB. Instead, the agency believed
there were insufficient data for immediate suspension of use of the fumigant
chemical.

This dynamic view of the decision-making process begs some hard
questions: What prevented FDA from banning DES immediately in 1971?
If subsequent evidence proved contradictory, the ban could have been
modified or lifted. What prevented EPA from immediately suspending
the use of EDB as a fumigant? Again, if subsequent data had shown
extremely low residues in foodstuffs, the ban could have been modified.
What prevented the medical community (and manufacturers of cigarettes)
from warning the public immediately in 1953 (and even earlier) of the
serious, legitimate evidence that cigarette smoking may cause lung cancer?
If further research had shown otherwise, a superseding statement of opinion
could have been issued.

Implicit in these examples is the assumption that an action taken now
cannot be rescinded—or, more precisely, that undoing an action is quite

costly. Thus, implicit in Passey's argument is the contention that it would be quitely costly for the public to recover from a false alarm about smoking and cancer. Implicit in Hutt's description is that the act of bypassing the normal hearing process on DES would have been a costly administrative and political error. In these instances, an unstated assumption of irreversibility creates a bias toward waiting.

The concept of irreversibility of decisions has not been considered in the literature on environmental policy making. Yet economists have made a number of attempts to spell out its consequences, especially in recent theoretical work in financial economics (Henry, 1974; Cukierman, 1980; Roberts and Weitzman, 1981; Baldwin, 1982; Bernanke, 1983; McDonald and Siegel, 1986; Majd and Pindyck, 1987).

In the economic models, a decision maker is assumed to be continuously faced with three types of choices: (1) invest in, (2) proceed with, or (3) abandon a hypothetical project. Investing, on the one hand, is a noncommittal action. It may accelerate the arrival of new information about a project's benefits and costs, but the project's ultimate fate remains undecided. On the other hand, the decisions to proceed with or to drop the project are assumed to be irreversible.

The assumption of irreversibility has a number of simple consequences in the economic models. In particular, conventional, static benefit-cost analysis is rendered misleading (Majd and Pindyck, 1987). Even if the expected benefits of a project exceed its expected costs at a particular point in time, the decision to proceed may be unwarranted. Instead, the decision criteria should be modified to take into account the benefits and costs of waiting for more information. The modified decision rule is to take action only when expected benefits exceed costs by a fixed, predetermined amount. (Strictly speaking, this rule is applicable only when the stochastic process that generates new information is stationary; see, for example, Roberts and Weitzman [1981].) Put differently, the expected net benefit of the project has to exceed an "option value" of waiting for more information.

These stylized, economic models of the wait-or-act decision have general application. The financial decision to proceed with or abandon a project is analogous to the public policy decision to approve or disapprove, let us say, a new drug application or cleanup technology. The financial decision to invest parallels the regulatory decision to send the drug or technology back for more study.

The critical issue in applying the economic models, however, is the validity of their assumption of irreversibility. It is counterproductive to jump to label an environmental regulation as irreversible until the sunken costs that must be expended to comply with the regulation are actually measured.

In conducting such an empirical inquiry, what is needed is a typology of sunken costs. As a preliminary scheme, I shall suggest three classes: (1) producer compliance costs, (2) consumer compliance costs, and (3) credibility costs. The first two categories reflect responses by producers and consumers, respectively, to environmental policy decisions. Thus, banning saccharin might result in a permanent and costly shutdown in saccharin-producing facilities. Prohibiting the use of DES as a livestock fattening agent might result in permanent and costly changes in the consumer diet. Credibility costs, the third category, arise because policy decisions are interdependent. Consumers' and producers' responses to environmental policies depend on the credibility of the policy-making entity. If the FDA banned saccharin or DES immediately and if the action turned out to be mistaken, then the agency's ability to enforce subsequent regulatory actions might be destroyed.

Still, we need to ask for hard evidence to ensure that capital in the saccharin industry was, indeed, nontransferable. We need to inquire whether consumers could go back to leaner meats if and when DES were reintroduced. We also need to ask whether the credibility costs of policy mistakes in reality all argue in favor of waiting.

WAITING AND SUNKEN COSTS

The argument in favor of regulatory delay, we have seen, hinges critically on the proposition that government intervention may impose irreversible, sunken costs on private agents. In this section, I suggest that the irreversibility argument can be turned upside down: waiting can have equally irreversible consequences.

When a potential environmental hazard is first recognized, its control may be amenable to partially reversible interventions (e.g., restrictions on access or use, product warning or labeling, pollution fees). If the hazard later becomes quite large, however, then such small-scale interventions may be ineffective, and only large-scale, irreversible interventions may be worth considering. Thus, the regulator who waits for more data runs the risk that only the most extreme, irreversible measures will be available in the end. Acid rain and toxic waste disposal may be good examples of the problem of increasingly narrow regulatory choices.

It is no accident of nature that the costs of effective intervention grow larger when regulatory agencies delay action. Private economic agents, especially business firms, have an incentive to make intervention costly. The longer the regulatory authority waits, the more "breathing time" firms may have to commit themselves to the suspect technology.

Diesel Emissions

Since the 1950s, the condensates from diesel fuel-burning engines have been known to cause cancer in laboratory animals. These particulate emissions are further known to contain carcinogenic polyaromatic hydrocarbons. Yet, there has been little sound epidemiological evidence available on the cancer risks of workers exposed to such emissions.

In the late 1970s, in the face of increasing pressures for fuel economy, American automobile manufacturers announced plans to convert 25 percent of their light-duty passenger car fleet from gasoline to diesel fuel-burning engines. The result of such a conversion would have been an increase in population exposures to particulate emissions by an estimated factor of 1,000. The auto makers' proposal stimulated new research into the combustion process and the physical chemistry of the particulate matter contained in diesel and other emissions. By 1979, EPA scientists determined that the organic solvent extracts of diesel particulates were highly mutagenic in the Ames mutagenicity assay. Directly mutagenic nitroaromatic compounds were identified as the likely culprits.

EPA lauched a major research program that included laboratory testing of fossil fuel combustion products. The carcinogenicity and mutagenicity of diesel and other emissions were confirmed in multiple laboratory models. Mathematical extrapolations suggested a small individual risk of cancer, but the estimated number of exposed persons was quite large. There was renewed interest in epidemiological studies of exposed workers but very little hard evidence available on the effects of emissions on humans. A study of London transport workers was negative, but was of sufficiently low power that some lung cancer risk from diesel emissions could not be excluded (Harris, 1983).

A scientific panel of the National Research Council could do no more than reiterate the substantial existing uncertainty about the health risks of the proposed diesel technology (National Research Council, 1981). Moreover, although the biological data base gradually became more refined, the uncertainty about population exposures grew. Changes in the relative prices of diesel and gasoline fuels, as well as unanticipated changes in consumer preferences, made the large-scale introduction of diesel passenger cars less likely. What is more, there were continued uncertainties about the feasibility of effective, low-cost particulate control technologies for disel engines.

In the face of all of these uncertainties, EPA proposed immediate particulate emission standards for diesel cars (at a level of 0.6 gram per mile). This action hardly settled the issue, for it remained unclear whether the proposed standards should remain in effect or whether they should be tightened in the future. At the time, a stricter standard (0.2 gram per mile) was contemplated. Even if particulate standards were to be tightened, however, the agency still needed to know when to impose them.

By the early 1980s, EPA could reasonably conclude that diesel emissions had at least the potential to cause cancer in humans. With virtually no solid epidemiological evidence, however, the agency could not draw definite conclusions about the extent of human cancer risk. From a purely scientific standpoint, the prudent decision was to wait for the results of newly commissioned epidemiological studies. Concrete results from such studies were expected within five years.

EPA's decision was not as simple as it might appear, however. The planned conversion to a diesel-driven auto fleet would require a major investment in a new engine technology. Auto makers could not simply modify the existing production technology for gasoline-burning engines. If diesels were to constitute as much as 18 percent of new car sales by 1990, investments on the order of $3–$4 billion would be required. Moreover, it was unclear whether auto makers might later be able to convert the diesel technology to the production of gasoline-burning engines. As the National Research Council reported, "[b]ased on the current state of knowledge, an irrevocable decision by the EPA . . . could run a danger of costly mistakes" (National Research Council, 1982).

Anyway, what did the agency really expect from the additional planned research? EPA could reasonably conjecture that by 1985, retrospective studies of workers exposed to diesel emissions might show an elevated risk of lung cancer. Such studies might bolster the case for regulation of diesel particulates. Still, the results of high-dose exposures in the workplace could not be simply extrapolated to low-dose ambient exposures from tail pipe emissions. Moreover, detailed laboratory studies of the composition and biological action of diesel particulate emissions still might not settle a key, lingering question: Did the apparently unique nitroaromatic constituents in the particulate extracts make diesel fumes a uniquely dangerous species of emissions?

What made EPA's regulatory dilemma so acute was not the laboratory discovery that diesel emissions were mutagenic, and not the paucity of direct, human evidence, but the announced intention of manufacturers to sink billions into a new diesel technology.

In fact, the agency was engaged in a prototypical regulatory game with the car makers. The longer EPA waited for new information, the further down the diesel road the car makers would be. The investment in diesel technology would not be instantaneous but gradual over a period of a decade or more. By the time EPA had sufficient information to satisfy the blue-ribbon scientific panels, the industry might have invested so much in diesel technology as to make tight emission controls too costly.

In this regulatory game, both EPA and the car makers knew the dilemma the agency might soon face. Hence, car makers had a strong incentive to accelerate their investments in diesel technology; that is, to build up their sunken costs as rapidly as possible. While EPA and some auto companies were conducting their own biological research, information on the likely pace of such research was common knowledge. On the other hand, the car makers possessed far more information on the irreversibility of investments in diesel production technology. In fact, EPA's lack of expertise in this area was perhaps its central difficulty in reaching a regulatory decision.

In the end, EPA stuck with its proposed emission controls, if only to avoid more drastic interventions later. As it turned out, however, the anticipated major demand for diesel cars never materialized, and the agency bought more time to wait for new data.

Cyanazine

To obtain registration for a pesticide under the Federal Insecticide, Fungicide, and Rodenticide Act or FIFRA (7 U.S.C. 136 et seq.), an applicant for registration must demonstrate, among other things, that the pesticide performs its intended function without causing "any unreasonable risk to man or the environment, taking into account the economic, social, and environmental costs and benefits of the use of any pesticide" (Section 2[bb]). EPA, the enforcing agency for the act, interprets this standard to require "a finding that the benefits of the use of the pesticide exceed the risks of use, when the pesticide is used in compliance with the terms and conditions of registration or in accordance with widespread and commonly recognized practice" (U.S. Environmental Protection Agency, 1988:795). If at any time EPA should determine that this benefit-cost standard has been violated, then the administrator may modify the conditions of registration or cancel the registration entirely.

In April 1985, EPA initiated a "special review" of all pesticide products containing the active ingredient cyanazine (U.S. Environmental Protection Agency, 1985). The review (formerly called the "Rebuttable Presumption Against Registration" or RPAR process) was instigated following the recent finding that cyanazine produced teratogenic and fetotoxic effects in laboratory animals. EPA proposed that a warning be added to the pesticide label concerning cyanazine's potential to cause birth defects in laboratory animals. Moreover, because the main route of occupational exposure was through skin contact, the product label was to specify that cyanazine's use was restricted to certified applicators or to persons under their supervision.

EPA was also concerned about groundwater contamination from agricultural uses of cyanazine. Preliminary monitoring studies had identified residues of cyanazine in a small percentage of sample wells from five states. Although most positive samples showed cyanazine concentrations of 0.2 part per billion (ppb), a small percentage showed levels close to 1 ppb. The agency thus noted:

> Cyanazine has the potential to move (leach) through the soil and contaminate ground water which may be used as drinking water. Cyanazine has been found in surface and ground water as a result of agricultural use. The Agency does not have the data necessary to assess the health risks associated with consuming drinking water which has been contaminated with cyanazine. (U.S. Environmental Protection Agency, 1985:14151)

Accordingly, the agency imposed labeling requirements that advised users not to apply cyanazine to highly permeable soils or to areas in which the water table was close to the surface. It also required registrants to conduct groundwater and surface water monitoring studies.

In a January 1987 review, the agency proposed a number of additional requirements for cyanazine registration, including the use of protective gloves, closed loading systems, and chemical-resistant aprons. The pesticide label was to include statements regarding the cleaning of protective gloves and separate laundering of protective clothing. In addition, the label was to state that cyanazine was classified for restrictive use because it "has caused birth defects in laboratory animals and has been found in ground water" (U.S. Environmental Protection Agency, 1987a:589).

By early 1988, however, new data suggested that cyanazine was not as serious a threat to groundwater as had been supposed. In particular, further sampling from 200 wells in hydrogeologically vulnerable areas revealed no detectable residues. The agency thus lifted its prior restriction on the spraying of cyanazine in areas in which the water table was high or the soil was highly permeable.

> As a result of newly generated monitoring data and the previously available data, the Agency no longer believes that cyanazine has significant ground water contamination potential. Therefore, EPA no longer believes that ground water contamination should be a reason for classifying cyanazine for Restricted Use. Therefore, all cyanazine labels will include a statement that cyanazine products have been classified for Restricted Use only because cyanazine has caused birth defects in laboratory animals. However, because some instances of contamination were reported in the earlier studies, the Agency believes the ground water advisory statement should remain on the label. (U.S. Environmental Protection Agency, 1988:795)

In the case of cyanazine, EPA altered its position several times as new evidence accumulated on the pesticide's potential toxicity and the routes of environmental exposure. The agency in fact reversed itself on the issue of groundwater contamination. The only clear effect of these multiple regulatory changes, however, was to alter the contents of the pesticide's warning label.

Ninety-six percent of the cyanazine produced in the United States was used as a herbicide on corn. About 3 percent was used on cotton, and less than 1 percent was used on sorghum and wheat. About 14–16 percent of the total U.S. corn acreage was treated with cyanazine in 1982. Several close substitutes for cyanazine were readily available, and there was little evidence that switching to these substitutes would be costly.

EPA was thus in a position to make a series of incremental changes in its regulation of cyanazine use without imposing large sunken costs on the private sector. Users of cyanazine were required to make investments in closed loading systems and protective equipment, but none of these investments was specific to a single chemical. Producers of cyanazine were required to reissue warning labels. In the absence of an outright ban on the use of cyanazine, however, the question of irreversible, cyanazine-specific investments did not arise.

Ethylene Dibromide

Table 1 traces scientific developments concerning ethylene dibromide (EDB) from 1910 to 1976. EDB was first used by producers of lead antiknock compounds for gasoline in the 1920s. By the late 1940s and early 1950s, the compound was widely employed as a fumigant of imported fruits and vegetables, grain, storage silos, and grain-milling machinery.

Data on EDB's acute and subacute toxicity go back to the early 20th century. The evidence on EDB arose from reports of accidental human exposure and from studies of ingestion, inhalation, and dermal exposure in various laboratory animals. By the mid-1960s, additional reports appeared on EDB's reproductive toxicity in farm animals. Still, residues of EDB remained essentially undetectable in the food supply.

In the early 1970s, two developments—the linking of EDB to mutagenicity and carcinogenicity and the improvement of the technology for detecting EDB—brought increased attention to and concern about the compound. In 1971, EDB was found to be a direct-acting mutagen in the Ames mutation assay. By 1974, the chemical's genotoxicity had been confirmed in other experimental systems. At this time, scientists were increasingly interested in the possible role of genotoxic events in the genesis of cancer. The finding that EDB was a mutagen stimulated whole-animal carcinogenicity studies by the National Cancer Institute (NCI).

NCI's preliminary results showed that EDB was carcinogenic when it was directly instilled into the stomachs of rodents. To be sure, there was concern that the NCI results were somehow artifactual because the

TABLE 1 Reported Scientific Developments Surrounding Ethylene Dibromide (EDB), 1910-1976

Year	Scientific Developments
1910	Marmetschke reports on the acute human toxicity of EDB after accidental administration.
1923	EDB is first produced on a commercial scale for sale to producers of lead antiknock compounds.
1925	Neifert reports the efficacy of EDB as a fumigant.
1927	Thomas and Yant report EDB absorption in toxic amounts through the skin of exposed workers; acute toxicity is reproduced in the laboratory by inhalation and dermal exposure in guinea pigs.
1928	Kochmann reports on subacute toxicity in a worker repeatedly exposed to EDB and confirms acute toxicity in laboratory rabbits and cats exposed by inhalation.
1929	Glaser and Firsch confirm acute toxicity of EDB in guinea pigs.
1938	Pflesser reports on acute toxicity in workers exposed to EDB.
1946	Aman reports acute toxicity of EDB by oral administration in rats and guinea pigs.
1950	EDB comes into widespread use for quarantine treatments of imported fruits and vegetables, control of interstate movement of insect pests, fumigation of grain, spot fumigation of milling machinery, and soil fumigation.
1952	Adams et al. report acute toxicity in workers exposed to fumigant mixtures of EDB, ethylene dichloride, and carbon tetrachloride. Rowe et al. report acute toxicity by oral instillation, dermal and eye contact, and inhalation in rats, guinea pigs, rabbits, mice, chickens, and monkeys.
1955	McCollester et al. report acute toxicity to albino rats of EDB and of fumigant mixtures containing EDB. Bondi et al. report decreased egg production and egg weight in hens fed grain fumigated with EDB.
1960	Olmstead reports case of acute toxicity after accidental oral ingestion of EDB capsules.
1965	Amir and Volcani publish initial report on spermicidal action of EDB in bulls given feed containing EDB.
1968	Alumot reports reductions in egg size and egg fertility in hens given EDB-fumigated feed.
1970	Edwards et al. report rapid absorption and wide organ distribution of EDB in mice.
1971	Ames reports direct mutagenicity of EDB in Salmonella revertant assay.
1972	Buselmaier et al. confirm direct mutagenicity of EDB in Salmonella revertant assay.

TABLE 1 (Continued)

Year	Scientific Developments
1973	Olson et al. report preliminary results of a National Cancer Institute (NCI) oral gavage study in Osborne-Mendel rats and B6C3F1 mice; squamous cell carcinomas of the stomach are observed in experimental animals near sites of application. Amir reports that the spermicidal effect of EDB in bulls results from direct action on spermatogenesis; the effects appear to be reversible.
1974	U.S. production of EDB reaches 330 million pounds, of which 200 million pounds are used in lead antiknock formulations. Brem et al. confirm the experimental mutagenicity and DNA-modifying effects of EDB. Vogel and Chandler confirm mutagenicity in Drosophila.
1975	Powers et al. report additonal results of NCI oral gavage study in rats and mice. EPA study shows gasoline station exposures to EDB in the range of 0.01 parts per billion and manufacturing site exposures in the range of 10 to 15 parts per billion.
1976	Plotnick and Conner confirm wide organ distribution of EDB in guinea pigs after experimental administration.

experimentally induced stomach cancers appeared near the site of EDB application. Still, the prospect of EDB's carcinogenicity changed the entire perspective on the chemical's risks. Many scientists believed that there was no dosage threshold for carcinogenicity. With EDB a potential carcinogenic contaminant of the food supply, many believed it was possible that even traces of residual pesticide were causing cancer in humans.

By 1975, an EPA study had shown detectable gasoline station exposures in the range of 0.01 ppb, and manufacturing site exposures in the range of 10–15 ppb. These findings heightened the concern over the long-term consequences of low-dose EDB exposures.

Table 2 continues the EDB chronology from 1977 to 1984. The table displays not only the salient scientific developments but also some key regulatory actions. It constitutes a preliminary attempt to show the timing of regulatory decisions in relation to the emergence of new scientific and economic information. Not shown in Table 2 are contemporaneous developments in the media and public opinion. As shown in the table, the scientific evidence on EDB's potential hazards continued to accumulate after 1977. Yet media coverage of EDB erupted only after the chemical was discovered in groundwater in Florida, Georgia, California, and Hawaii in 1983. In that year, EPA issued an emergency suspension of soil fumigation using EDB. In the following year, the agency announced the suspension of all further use of EDB in the production of grain products (Russell and Gruber, 1987).

TABLE 2 Scientific and Regulatory Developments Surrounding Ethylene Dibromide (EDB), 1977-1984

Year	Scientific Developments	Regulatory Developments
1977	Ott and Scharmweber report on 156 Dow Chemical employees in two EDB production facilities; no significant increase in mortality or cancer is found. International Agency for Research on Cancer (IARC) classifies EDB as an experimental mutagen and animal carcinogen. Hunt, in the Great Lakes Chemical Corp. submission to the Occupational Safety and Health Administration (OSHA), reports current worker exposure during EDB fumigant application to be in the range of 60-520 parts per billion (ppb), depending on adherence to label directions. In a risk assessment based on the National Cancer Institute (NCI) oral gavage study, EPA's Carcinogen Assessment Group (CAG) predicts almost 100 percent lifetime incidence of cancer from a 40-year exposure to 400 ppb of EDB.	The National Institute on Occupational Safety and Health (NIOSH) reviews data on pharmacokinetics, metabolism, acute and chronic toxicity, reproductive effects, and carcinogenicity of EDB. OSHA recommends tightening of the standard for occupational exposure to 20 parts per million (ppm) time-weighted average (TWA) of EDB. The Environmental Protection Agency (EPA) issues its Position Document no. 1 on EDB and initiates the "Rebuttable Presumption Against Registration" (RPAR) process of EDB under the Federal Insecticide, Fungicide and Rodenticide Act (FIFRA).
1978	Rausch, in a Dow Chemical submission to OSHA, reports on current and historical occupational exposures to EDB; exposures were 1-24 ppm in 1949 and 1952 and less than 5 ppm in 1971 and 1972. Ter Haar, in an Ethyl Corp. submission to OSHA, issues a preliminary report on mortality and reproductive function in workers exposed from 3 months to 10 years. At dosages ranging from less than 0.15 ppm to 4.5 ppm, no elevation in death rates was detected. Sperm	

TABLE 2   (Continued)

Year	Scientific Developments	Regulatory Developments
	counts compared favorably to those of the general population. Trend of sperm counts in relation to EDB exposure is found but is of questionable significance. NCI publishes the results of an oral gavage study in rats and mice.	
1979	Van Duuren et al. report on skin painting study in Ha:ICR Swiss Webster mice; EDB is found to be carcinogenic. Wong et al. report a retrospective evaluation of reproductive performance of workers exposed to EDB; no effects are seen in three of four plants. Plotnick et al. publish preliminary results of a NIOSH inhalation study in Sprague-Dawley rats. Ramsey et al. report that CAG's 1977 risk assessment would predict 54-85 cancer cases among 156 exposed Dow Chemical employees in the Ott-Scharmweber study; 8 cases were actually observed.	EPA cancels registration of the soil fumigant dibromochloropropane (DBCP), probably resulting in increased use of EDB.
1980	Ott et al. publish a follow-up report on the Dow Chemical cohort; the findings are inconclusive due to small cohort size, incomplete exposure data, incomplete follow-up, and confounding with other chemical exposures (arsenicals). Terr Haar publishes a follow-up report on a cohort of 53 employees exposed to EDB; the cohort is too small to assess the cancer risk. An NCI inhalation study on Fisher-344 rats and B6C3F1	EPA issues Position Document no. 2/3 (Notice of Preliminary Determination Concluding the Rebuttable Presumption Against Registration of EDB); it proposes to continue registration of EDB for preplant soil fumigation but wishes to cancel EDB registrations for fumigations of stored grains and spot fumigation of grain-milling machinery. The agency also proposes canceling postharvest

TABLE 2 Continued

Year	Scientific Developments	Regulatory Developments
	mice is submitted for internal peer review; EDB is found to be carcinogenic. Wong et al. publish intermediate results of a NIOSH inhalation study in rats; EDB is found to be carcinogenic. The American Conference of Governmental Industrial Hygienists classifies EDB as a suspect human carcinogen. An EPA internal review estimates the probable residue level for EDB in wheat bread derived from grain fumigated after harvest with EDB to be 0.07 ppb; the "realistic worst case" residue is estimated to be 31 ppb. EPA's CAG issues its cancer risk assessment, based on a one-hit mathematical model; the estimated lifetime cancer risk of the dietary burden of EDB is estimated at 3.3 per 10,000. EPA commissions a groundwater contamination study by the California Department of Food and Agriculture (CDFA).	fumigation of fruits and vegetables by July 1, 1983. EPA requires soil fumigant registrants to conduct groundwater contamination studies. The U.S. Supreme Court requires that a "significant risk" be adduced to justify OSHA regulations (Industrial Union Department v. American Petroleum Institute et al.).
1981	Publication of the the final results of NIOSH inhalation study in rats: EDB is found to be carcinogenic. EDB is used in California to fight the Mediterranean fruit fly. Maddy et al. (CDFA) estimate EDB residues of up to 57 ppb in edible portions of fumigated citrus fruits. Raines and Holder find an average EDB residue of 35.7 ppb in biscuits, contrary to early EPA estimates of 0.07 ppb;	The State of California (Cal/OSHA) issues temporary emergency standard of 130 ppb. OSHA issues Advance Notice of Proposed Rulemaking regarding occupational exposure to EDB, proposing a reduction of the standard from 20 ppm to 15 ppb and requesting comments on quantitative risk assessment (Federal Register, December 18).

TABLE 2 Continued

Year	Scientific Developments	Regulatory Developments
	reported levels in flour range from nondetectable to 4.2 ppm.	
1982	Publication (in March) of the final results of the NCI inhalation study in rats and mice: EDB is found to be carcinogenic. EPA scientists are notified (in June) that three wells in Seminole County, Georgia, are contaminated with EDB levels as high as 100 ppb. SRI International publishes a NIOSH-commissioned risk assessment based on NCI and NIOSH inhalation studies in rats and mice (June); chronic exposure to 130 ppb is predicted to yield 4-26 percent lifetime human cancer risk. CDFA (June 2) revises estimates of EDB residues in fumigated citrus fruits up to 210-880 ppb. Wade and Sakura report two acute lethal reactions among workers exposed to EDB.	OSHA interprets the Supreme Court ruling as permitting mathematical risk assessment in support of agency regulations (Federal Register, April 9). Cal/OSHA's emergency standard of 15 ppb is rejected by California Office of Administrative Law; California adopts as a permanent regulation a standard of 130 ppb.
1983	The National Toxicology Program reports that inhalation of EDB (10-40 ppm) in Fisher 344 rats produced testicular degeneration. An EPA-commissioned study of groundwater contamination by CDFA issues its preliminary report (Spring), finding EDB at concentrations between 0.1 and 31 ppb in the soil at depths greater than 20 feet, moving down to groundwater. A follow-up report (June) reveals groundwater levels between 0.02-5 ppb in 16 counties in 4	EPA issues Position Document no. 4 (September 27), with revisions in its mathematical risk assessment methodology. EPA issues an emergency suspension of its soil fumigation with EDB; it gives notice (September 28) of intent to cancel registration of EDB as a grain and fruit fumigant under the "unreasonable hazard" standard of FIFRA. EDB use in fumigation is to be eliminated by 1986. The state of Florida issues emergency regulations restricting EDB in

TABLE 2 Continued

Year	Scientific Developments	Regulatory Developments
	states. A new EPA risk analysis is issued as part of Position Document no. 4. The original one-hit model of Position Document no. 2/3 is modified to include "Weibull timing." The estimated average EDB content of grains is revised upward markedly to 31 ppb. CAG's new estimate of lifetime cancer risk from a dietary burden of EDB is 3.3 per 1,000, based on lifetime consumption of current levels of EDB in grain products.	uncooked grain products to 1 ppb (level of detection).
1984	Grocery Manufacturers of America (GMA), modifying the Rains and Holder (1981) detection methodology, find that 79 percent of ready-to-eat, grain-derived products contain EDB levels below 1 ppb; GMA also reports on the disappearance of EDB through cooking raw grain products. Environ Corporation, under GMA sponsorship, issues (January 20) risk assessment of exposures to EDB residues in consumable grain products, based on NCI oral gavage assay and assumptions of no further grain fumigation and of the depletion of EDB in grain stores by 1986; the upper limit of lifetime cancer is estimated to be 1 in 4 million. Temple, Barker & Sloane, Inc., and Economic Perspectives, Inc. issue an economic analysis of the impacts of immediate removal of EDB from the food supply; if 50-60 percent of stored	EPA announces (February 3) immediate suspension of further use of EDB in the production of grain products and recommends guidelines to states for acceptable levels of EDB in foods, including 900 ppb in raw grain products, 150 ppb in processed products requiring further cooking, and 30 ppb in ready-to-eat foods. The Massachusetts Department of Public Health recommends (February 6) emergency regulation at 10 ppb for all food products, with transition in 30 days to 1 ppb. ("The Department's position is that the only safe level of exposure to a carcinogen is one that is zero or near zero. The Department therefore believes that it is appropriate to move rapidly to levels of EDB in food of less than 1 ppb.")

TABLE 2 Continued

Year	Scientific Developments	Regulatory Developments
	grains and 67 percent of grain products were immediately restricted from any use, they conclude, grain prices would nearly double, with consumer expenditure increases of $35 billion and grocery manufacturer losses of $2.8 billion in inventories.	

EPA's suspensions of the use of EDB in 1983 and 1984 were not the first regulatory actions taken with respect to the pesticide. Nor did 1983 see the first instance of damning evidence on EDB. The question arises: What exactly happened between 1977 and 1983?

By 1977 the International Agency for Research on Cancer had already classified EDB as an animal carcinogen and mutagen. A review by the National Institute on Occupational Safety and Health (NIOSH) noted that EDB was able to interact chemically with deoxyribonucleic acid (DNA), the basic genetic material. Still, EDB had thus far been found to be carcinogenic in only one incomplete animal experiment. Moreover, attempts to identify elevated cancer rates among EDB-exposed workers were unsuccesful. If EDB in fact posed a cancer threat at low doses, the magnitude of the cancer risk remained uncertain.

In the face of this uncertainty, the Occupational Safety and Health Administration (OSHA) proposed a tightening of its EDB exposure standard for workers. EPA, in parallel, began a special RPAR review under FIFRA. The linchpin of EPA's regulatory analysis was a risk assessment, performed by its Carcinogen Assessment Group (CAG).

CAG's initial risk assessment proved to be problematic. The initial dosages of EDB in the NCI oral gavage study—on which CAG relied—proved to be too toxic, so the dosage schedule had to be reduced in the middle of the experiment. This changing dosage schedule complicated CAG's attempts to extrapolate from high-dose to low-dose effects and risks. The CAG analysis also predicted a substantial cancer risk from long-term EDB exposures at the levels seen among chemical workers; limited surveys of EDB-exposed workers, however, showed no evidence of a significant cancer increase.

Yet by 1979, additional laboratory studies had confirmed EDB's carcinogenicity. The chemical caused cancers by skin painting in mice, and a NIOSH-sponsored study showed cancers by inhalation in rats. By 1980, EDB was found to be carcinogenic in a separate NCI-sponsored inhalation study of rats and mice. In that year, the American Conference of Governmental Industrial Hygienists also classified EDB as a suspect human carcinogen.

EPA's special RPAR review continued in 1980. An internal study estimated the probable residue level for EDB in wheat bread made from fumigated grain to be less than 0.1 ppb, with a realistic worst-case residue of 31 ppb. Based on such exposure estimates and extrapolating from the original NCI oral gavage experiment in rodents, CAG projected a 0.03 percent increased lifetime cancer risk owing to the dietary burden of EDB. The agency proposed cancellation of EDB's use as a fumigant of stored grains, milling machinery, and fruits and vegetables by mid-1983. It also ordered studies of potential groundwater contamination.

By 1981, new measurements of EDB residues in fruit and grain products suggested that previous estimates had been misleading. One study found EDB residues of 36 ppb in biscuits. Another found 57 ppb in the edible portions of fumigated fruits. Concurrently, OSHA proposed further tightening of the occupational standard for EDB exposure; California imposed a temporary emergency occupational standard.

By 1982, EDB levels as high as 100 ppb had been found in three wells in Georgia. The California Department of Food and Agriculture (CFDA) estimated that fumigated citrus fruits contained EDB residues of up to 210–880 ppb. By spring 1983, CFDA had found EDB concentrations of 0.1–31 ppb at depths greater than 20 feet. By June 1983, EDB had been detected at levels of 0.02–5 ppb in 16 counties.

EPA moved in September 1983 to suspend soil fumigation immediately. Based on the new exposure data, as well as a reanalysis of the NCI oral gavage experiment, CAG revised the estimated lifetime risk from dietary EDB to 0.3 percent. In February 1984, the agency suspended further use of EDB in the production of grain products, although it did not order an immediate ban on the sale of all EDB-containing products. Instead, it issued recommended guidelines to the states for acceptable levels of EDB in currently marketed foods.

Why did EPA wait six years (from its initial review in 1977 until its emergency suspension in 1983) to take action on EDB? The evidence of EDB's toxicity was long-standing: its mutagenicity was established in 1971, and its carcinogenicity was reported by 1977. Although the initial NCI study required confirmation, independent findings of carcinogenicity were available by 1979. Initially, EDB was thought to be virtually undetectable in the food supply; yet contrary evidence was available by 1981. Groundwater contamination was an issue as early as 1980, when EPA commissioned a study by CDFA. Residues were found in wells as early as June 1982.

Perhaps it is unfair to juxtapose EPA's regulation of cyanazine during 1985–1988 with the agency's drawn-out response to EDB during 1977–1984. By the mid-1980s, the agency had improved its handling of procedural and notification burdens built into FIFRA, which was enacted in 1972. Still, the cyanazine case shows the agency moving quickly in incremental, reversible steps to establish warning labels and restrictions on use. In the case of EDB, the agency essentially found itself having to ban the pesticide late in the game, years after other federal and state agencies had moved on the problem. Had EPA accelerated the information-gathering process, especially in the measurement of food residues and groundwater contamination, less extreme measures might have been necessary.

By 1984, the sunken investment in EDB had become enormous: $29 billion in grain stocks and $4.3 billion in manufacturer and retail inventories of grain products and baked goods. It was likely that between 50 and 60 percent of stored grains and grain products contained detectable levels of EDB. Commingling of grains during storage, transport, and manufacture raised the possibility that nearly all such products had detectable levels of the chemical (Temple, Barker and Sloan, Inc., and Economic Perspectives, Inc., 1984). Immediate removal of EDB-containing foods would have been quite costly. In the end, EPA chose an intermediate course: suspension of use of the compound without confiscation of existing stocks of potentially EDB-contaminated food.

REGULATION AS RESEARCH

Scientists and policy makers may recommend delaying regulatory action until they see the results of current research. Yet the need to perform more research does not preclude concurrent regulatory intervention. EPA imposed a groundwater advisory on cyanazine's label even as it sought further testing of pesticide residues. The agency imposed a standard on particulate emissions from diesel-powered cars even as it awaited the results of epidemiological studies on diesel workers. Although EPA did not restrict EDB until 1983, earlier action should not have barred further toxicological and exposure studies.

In fact, there is no clear dividing line between regulatory intervention and research. The reason is that knowledge can be gained from the experience of regulatory intervention. In some instances, the best way to assess the benefits and costs of regulation is to regulate and see what happens. By contrast, further delay may bring little or nothing in the way of new information.

The nation's experience with environmental controls may provide the best source of information—and sometimes the only source of information—on the costs of complying with even stricter controls. At issue here is whether the public or private sectors are best suited to perform the necessary research on new control technologies. When the development of new controls entails highly specialized or proprietary knowledge, it may be impractical for regulatory agencies to fund public research into cleanup technologies. Instead, the most effective way to instigate the necessary research is to impose environmental controls, thus changing the incentives of private firms.

Conversely, experience with regulatory controls may be the best or only means of assessing the benefits of environmental regulation. Laboratory experiments can measure small-scale individual effects, whereas environmental controls operate on a large scale. Thus, laboratory experiments and meteorologic modeling can offer only imprecise gauges of the aggregate effect on acid rain of curbing sulfur oxide emissions. Measurement of individual tail pipe emissions, in combination with dispersion modeling, may be inadequate to predict the aggregate effect of installing auto pollution control devices.

The main point is that small-scale "micro" models and experiments may be inadequate to understand or predict the "macro" consequences of large-scale policy interventions (Harris, 1985). At best, basic research and data acquisition can only disentangle individual mechanisms; they cannot by themselves show the interaction of multiple mechanisms of environmental damage and multiple routes of toxic exposure. The only way to assess such large-scale effects is by natural experiments; that is, by regulatory intervention.

CHLOROFLUOROCARBONS

In 1974, Molina and Rowland proposed that long-lived, stable chlorofluorocarbons (CFCs) could slowly migrate to the stratosphere, where they would release chlorine following contact with high levels of radiation. The resultant free chlorine could in turn act as a catalyst to break apart ozone molecules. Thus, CFCs might be steadily depleting the stratospheric layer of ozone, the shield that stops ultraviolet-B radiation from penetrating to the earth's surface.

The ozone depletion hypothesis was taken seriously by the scientific community, and early work on the topic includes a 1976 report by the National Academy of Sciences. In 1977, Congress amended the Clean Air Act (42 U.S.C. 7457[b]) and authorized EPA's administrator to issue regulations for controlling substances or activities "which in his judgment may reasonably be anticipated to affect the stratosphere, especially ozone in the stratosphere, if such effect in the stratosphere may reasonably be anticipated to endanger public health or welfare. Such regulations shall take into account the feasibility and the costs of achieving such control." The statutory language permitted EPA to act in the face of scientific uncertainty (U.S. Environmental Protection Agency, 1987b).

In 1978, EPA and the U.S. Food and Drug Administration moved to ban the use of CFCs as aerosol propellants in all but "essential applications." During the early 1970s, aerosol propellants constituted about 50 percent of total CFC use in the United States. Thereafter, CFC use in propellants declined markedly.

Largely in response to a series of National Research Council studies in the late 1970s, in 1980, EPA issued an Advance Notice of Proposed Rulemaking under the Clean Air Act. The notice proposed that the production of certain CFCs be frozen and suggested the possible use of marketable permits to allocate CFC production among various industries.

In the early 1980s, however, new data and models suggested that many other factors contributed to ozone depletion in the stratosphere. Carbon dioxide and methane, two atmospheric gases that have been increasing in concentration in recent years, appeared to buffer the ozone-depleting effects of CFCs. Moreover, although CFCs continued to be used as foam-blowing agents, refrigerants, and solvents, the decline in CFC aerosol propellant use resulted in a leveling off of worldwide CFC production.

Beginning in about 1983, the demand for nonaerosol uses of CFCs accelerated. Total production expanded to such a point that it now exceeds 1974 levels. Levels of CFC-11 (primarily used as a foam-blowing agent) and CFC-12 (primarily used as a refrigerant) are now rising at 5 percent annually, while CFC-113 (mainly used as a solvent for electronics and metal cleaning) has risen an estimated 10 percent annually. Moreover, there have been increases in demand for certain brominated compounds that are also thought to deplete stratospheric ozone (e.g., Halon-1211, which is used in specialized firefighting applications). These changes have been paralleled by continued increases in carbon dioxide and methane.

In 1985, the World Meteorological Organization (WMO) conducted a review of all ground- and satellite-based atmospheric ozone measurements to date. WMO concluded that ozone levels in the upper atmosphere had in fact decreased by 0.2–0.3 percent annually during the 1970s. Moreover, these decreases were offset by increases in ozone in the lower atmosphere, so that the total "column" ozone had remained unchanged.

In May 1985, however, Farman, Gardiner, and Shanklin reported that ozone levels in Antarctica, which were measured during the months of September to November, had declined by 40 percent since 1957, with most of the decline occurring since the mid-1970s. The discovery of this Antarctic ozone hole was completely unexpected; a 40 percent decline was not predicted by current atmospheric models of ozone depletion. By 1987, additional measurements of a key compound—chlorine monoxide—suggested that anomalous chlorine chemistry may have played a role in the development of the Antarctic hole. Such findings left open the possibility that seasonal declines in ozone above Antarctica were idiosyncratic and not reflective of global chemistry. Still, researchers have yet to determine the exact mechanisms responsible for the high levels of chlorine monoxide in the Antarctic hole and whether such unknown mechanisms are, indeed, unique to Antarctica.

Moreover, recently published evidence (Kerr, 1987) has challenged the conclusion that total column ozone is stable. Ground-based and satellite measurements now suggest a 3–5 percent annual decline during the 1980s. As in the case of the Antarctic ozone hole, these measurements fall outside of the uncertainty bounds computed from current atmospheric models, which predict that column ozone should not have decined by even 1 percent. A review of the newer data has now been instituted by the National Aeronautics and Space Administration and the National Oceanographic and Atmospheric Administration.

Why did the models fail to predict the 1987 results? One possibility is that the results are artifactual (e.g., misinterpreted satellite measurements). Another is that the models have failed to consider adequately the solar cycle or volcanic activity. Still, the main problem is that current models, which now include approximately 50 chemical species and simulate over 140 different reactions, may not be able to replicate atmospheric chemistry accurately. Have they failed to predict the limits by which the lower atmosphere can compensate for stratospheric ozone losses? Have they failed to predict the buffering effects of carbon dioxide and methane? Are estimates of the half-lives of certain CFCs (75 years for CFC-11 and 110 years for CFC-12) inaccurate?

On September 16, 1987, the United States and 23 other nations signed the Montreal Protocol on Substances That Deplete the Ozone Layer. The agreement set forth a timetable for reducing specified ozone-depleting chemicals, including a freeze on production at 1986 levels, followed by reductions during the 1990s. EPA, in anticipation of U.S. ratification of the Montreal Protocol, has already mandated the reporting of 1986 production, imports, and exports by American firms (U.S. Environmental Protection Agency, 1987b).

Formal benefit-cost analysis of CFC regulation is a formidable task. Models are needed to estimate the future decline in stratospheric ozone levels; the possible compensating increase in lower atmospheric ozone levels; the potential adverse effects of changes in atmospheric ozone, including increased incidence of skin cancers and cataracts, damage to aquatic organisms, and accelerated weathering of outdoor plastics; and the overall effects of global warming. In addition, the economic dislocation resulting from restrictions of CFCs and halons must be determined, including losses in refrigeration, foam production, cleaning of electrical equipment, and firefighting applications.

Still other information is also required. Can regulations really wait for better data and models on atmospheric chemistry ozone depletion? What will be the future evolution of such scientific information? Will implementation of CFC and halon controls now provide a critical source of data in understanding the ozone problem?

Regulation of CFCs and halons is hardly an all-or-none proposition. Should the Montreal Protocol go into force, and should the United States ratify it, EPA will be required to implement the 1986 production-level freeze and the planned reductions for the 1990s. The agency currently proposes to use a system of marketable licenses. Production or use charges are also under consideration. It is unlikely that EPA can project the consequences of these proposed regulatory schemes. Accordingly, in choosing which scheme to adopt, the agency needs to ask what near-term interventions are likely to provide information about future regulatory designs.

CONCLUDING COMMENTS

In environmental decision making, inconclusive scientific evidence is a commonplace occurrence. Still, regulatory agencies continue to make decisions in the face of such uncertainty.

In evaluating regulatory choices, it is hardly enough to assess the static benefits and costs of each regulatory option. Instead, regulatory agencies need to solve the problem of timing, which means assessing the benefits and costs of intervening now versus intervening later.

To attack the problem of timing, I have suggested that regulatory agencies ask two types of questions: Will we be able to take back the regulatory action? Will intervention be informative about future regulatory choices?

Environmental regulation takes many forms: requiring private firms to conduct studies or report data, suspending some uses of a chemical while permitting others, mandating or changing warning labels, issuing emergency suspensions, and scheduling phaseouts. In general, my analysis points toward a style of regulation in which agencies take small, incremental regulatory steps at the early stages of a problem. These small steps would be designed to impose minimal sunken investments in compliance, yet provide essential information on the uncertain benefits and costs of intervention.

The supporting evidence for the success of this style of regulation, however, has been largely anecdotal. I have cited a few possibly unrepresentative examples. To assess the results of past environmental decisions and to formulate guides for future choices will require a much wider array of case studies.

Still, I see broad application of the idea that environmental decision makers often wait too long to take action in the face of uncertainty. The reasons for delaying action, I suggest, are at best poorly articulated. Assertions that proof is not yet available, or that intervention will distract attention from more fundamental causes, or that the public will be needlessly alarmed, should be subject to more careful scrutiny. The refrain that "we need more research before we can act" likewise needs to be questioned. It is unfair to state the problem as "regulation versus research" when the main issues are, instead, the synergies between regulation and research.

REFERENCES

Baldwin, Carliss Y.
> 1982   Optimal sequential investment when capital is not readily reversible. *Journal of Finance* 37:763–782.

Bernanke, Ben
> 1983   Irreversibility, uncertainty, and cyclical investment. *Quarterly Journal of Economics* 98:85–106.

Cukierman, Alex
> 1980   The effects of uncertainty on investment under risk neutrality with endogenous information. *Journal of Political Economy* 88:462–475.

Harris, Jeffrey E.
> 1983   Diesel emissions and lung cancer. *Risk Analysis* 3:83–100.
> 1985   Macro-experiments versus micro-experiments for health policy. Pp. 145–185 in J. Hausman and D. Wise, eds., *Social Experimentation*. Chicago: University of Chicago Press.

Henry, Claude
> 1974   Investment decisions under uncertainty: The irreversibility effect. *American Economic Review* 64:289–322.

Hutt, Peter B.
> 1977   Question and answer session: FDA-diethylstilbestrol panel. Pp. 1675–1682 in H. H. Hiatt, J. D. Watson, and J. A. Winsten, eds., *Origins of Human Cancer. Book C, Human Risk Assessment*. Cold Spring Harbor, N.Y.: Cold Spring Harbor Laboratory.

Isselbacher, Kurt J.
> 1977   Statement of Kurt J. Isselbacher, M.D. Pp. 449–453 in *Propsed Saccharin Ban—Oversight*. Hearings before the Subcommittee on Health and the Environment, Committee on Interstate and Foreign Commerce, House of Representatives, U.S. Congress, March 21 and 22. Serial Number 95-8. Washington, D.C.: Government Printing Office.

Kerr, R. A.
> 1987   Has stratospheric ozone started to disappear? *Science* 237:131–132.

Majd, Saman, and Robert S. Pindyck
> 1987   Time to build, option value, and investment decisions. *Journal of Financial Economics* 18:7–27.

McDonald, Robert, and Daniel Siegel
> 1986   The value of waiting to invest. *Quarterly Journal of Economics* 101:707–727.

National Research Council
    1981   *Health Effects of Exposure to Diesel Exhaust.* The Report of the Health Effects
            Panel of the Diesel Impacts Study Committee. Washington, D.C.: National
            Academy Press.
    1982   *Diesel Cars. Benefits, Risks, and Public Policy.* Final Report of the Diesel
            Impacts Study Committee. Washington, D.C.: National Academy Press.
Passey, R. D.
    1953   Smoking and lung cancer (letter). *British Medical Journal* i(February 14):399.
Roberts, Kevin, and Martin L. Weitzman
    1981   Funding criteria for research, development, and exploration projects. *Economet-
            rica* 49:1261–1288.
Russell, Milton, and Michael Gruber
    1987   Risk assessment in environmental policy-making. *Science* 236:286–295.
Temple, Barker & Sloane, Inc., and Economic Perspectives, Inc.
    1984   *The Economics of Immediate EDB Removal.* Lexington, Mass.: Temple, Barker
            & Sloane.
U.S. Environmental Protection Agency
    1985   Special review of certain pesticide products; cyanazine. *Federal Register* 50(Apr.
            10):14151.
    1987a   Preliminary determination to cancel registrations of cyanazine products unless
            the terms and conditions of the registration are modified; availability of technical
            support document and draft notice of intent to cancel. *Federal Register* 52(Jan.
            7):589.
    1987b   Protection of stratospheric ozone; proposed rule. *Federal Register* 52(Dec.
            14):47489.
    1988   Cyanazine; intent to cancel registrations; denial of applications for registration;
            conclusion of special review. *Federal Register* 53(Jan. 13):795.
U.S. Environmental Protection Agency, Office of Policy Analysis
    1985   *EDB, A Case Study in the Communication of Health Risks.* Washington, D.C.:
            U.S. Environmental Protection Agency.

7

Choice Under Uncertainty: Problems Solved and Unsolved

MARK J. MACHINA

Fifteen years ago, the theory of choice under uncertainty could be considered one of the "success stories" of economic analysis: it rested on solid axiomatic foundations;[1] it had seen important breakthroughs in the analytics of risk and risk aversion and their applications to economic issues;[2] and it stood ready to provide the theoretical underpinnings for the newly emerging "information revolution" in economics.[3] Today, choice under uncertainty is a field in flux: the standard theory and, implicitly, its public policy implications are being challenged on several grounds from both within and outside the field of economics. The nature of these challenges, and of economists' responses to them, is the topic of this paper.

The following section provides a brief but self-contained description of the economist's canonical model of individual choice under uncertainty, the *expected utility* model of preferences over lotteries. I shall describe this model from two different perspectives. The first perspective is the most familiar and has traditionally been the most useful for addressing standard economic questions. However, the second, more modern perspective will be the most useful for illustrating some of the problems that have beset this model, as well as some of the proposed responses.

Each of the following sections is devoted to one of these problems. All are important; some are more completely "solved" than others. In each

---

Mark J. Machina is professor in the Department of Economics at the University of California, San Diego.

[1] See, for example, von Neumann and Morgenstern (1947), Marschak (1950), and Savage (1954).

[2] See, for example, Arrow (1963, 1974), Pratt (1964) and Rothschild and Stiglitz (1970, 1971). For surveys of applications, see Lippman and McCall (1981) and Hey (1979).

[3] See, for example, Akerlof (1970) and Spence and Zeckhauser (1971). For overviews of the subsequent development of this area, see Stiglitz (1975, 1985).

case, I begin with a specific example or description of the phenomenon in question. I then review the empirical evidence regarding the uniformity and extent of the phenomenon. Finally, I shall report on how these findings have changed, or are likely to change, or should change, the way economists view and model private and public decisions under uncertainty. On this last topic, the disclaimer that "my opinions are my own" has more than the usual significance.

THE EXPECTED UTILITY MODEL

The Classical Perspective: Cardinal Utility and Attitudes Toward Risk

In light of current trends toward generalizing this model, it is useful to note that the expected utility hypothesis was *itself* first proposed as an alternative to an earlier, more restrictive theory of risk-bearing. During the development of modern probability theory in the 17th century, such mathematicians as Blaise Pascal and Pierre de Fermat assumed that the attractiveness of a gamble offering the payoffs $(x_1, ..., x_n)$ with probabilities $(p_1, ..., p_n)$ was given by its *expected value* $\bar{x}$ (i.e., the weighted average of the payoffs where each payoff is multiplied by its associated probability, so that $\bar{x} = x_1 p_1 + ... + x_n p_n$). The fact that individuals consider more than just expected value, however, was dramatically illustrated by an example posed by Nicholas Bernoulli in 1728 and now known as the *St. Petersburg Paradox*:

> Suppose someone offers to toss a fair coin repeatedly until it comes up heads, and to pay you $1 if this happens on the first toss, $2 if it takes two tosses to land a head, $4 if it takes three tosses, $8 if it takes four tosses, and so on. What is the largest sure payment you would be willing to forgo in order to undertake a *single* play of this game?

Because this gamble offers a 1/2 chance of winning $1, a 1/4 chance of winning $2, and so forth, its expected value is (1/2)$1 + (1/4)$2 + (1/8)$4 + . . . = $1/2 + $1/2 + $1/2 + ... = ∞; thus, it should be preferred to *any* finite sure gain. However, it is clear that few individuals would forgo more than a moderate amount for a one-shot play. Although the unlimited financial backing needed to actually make this offer is somewhat unrealistic, it is not essential for making the point: agreeing to limit the game to at most a million tosses will still lead to a striking discrepancy between a typical individual's valuation of the modified gamble and its expected value of $500,000.

The resolution of this paradox was proposed independently by Gabriel Cramer and Nicholas's cousin Daniel Bernoulli.[4] Arguing that a gain of $2,000 was not necessarily "worth" twice as much as a gain of $1,000, they hypothesized that individuals possess what is now termed a *von Neumann-Morgenstern utility of wealth function* $U(\cdot)$. Rather than evaluating gambles on the basis of their expected value $\bar{x} = x_1 p_1 + \ldots + x_n p_n$, individuals will evaluate them on the basis of their expected utility $\bar{u} = U(x_1)p_1 + \ldots + U(x_n)p_n$ This value is calculated by weighting the utility of each possible outcome by its associated probability, and it can therefore incorporate the fact that successive increments to wealth may yield successively diminishing increments to utility. Thus, if utility took the logarithmic form $U(x) = ln(x)$ (which exhibits this property of diminishing increments) and the individual's wealth at the start of the game were, let us say, $50,000, the sure gain that would yield just as much utility as taking this gamble (i.e., the individual's *certainty equivalent* of the gamble), would be about $9, even though the gamble has an infinite expected value.[5]

Although it shares the name "utility," this function $U(\cdot)$ is quite distinct from the ordinal utility function of standard consumer theory. Although the latter can be subjected to any monotonic transformation, a von Neumann-Morgenstern utility function is cardinal in that it can only be subjected to transformations that change the origin point or the scale (or both) of the vertical axis, but do not affect the "shape" of the function. The ability to choose the origin and scale factor is often exploited to *normalize* the utility function—for example, to set $U(0) = 0$ and $U(M) = 1$ for some large value M.

To see how this shape determines risk attitudes, let us consider Figures 1a and 1b. The monotonicity of the curves in each figure reflects the property of stochastic dominance preference, by which one lottery is said to *stochastically dominate* another if it can be obtained from it by shifting probability from lower to higher outcome levels.[6] Stochastic dominance preference is thus the probabilistic extension of the attitude that "more is better."

Consider a gamble offering a 2/3 chance of a wealth level of x' and a 1/3 chance of a wealth levels of x''. The amount $\bar{x} = (2/3)x' + (1/3)x''$ in the figures gives the expected value of this gamble; $\bar{U}_a = (2/3)U_a(x') +$

---

[4] Bernoulli (1738). For a historical overview of the St. Petersburg paradox and its impact, see Samuelson (1977).

[5] Algebraically, the certainty equivalent of the Petersburg gamble is given by the value ξ that solves $U(W+\xi) = (1/2)U(W+1) + (1/4)U(W+2) + (1/8)U(W+4) + \ldots$, where W denotes the individual's initial wealth (i.e., wealth going into the gamble).

[6] Thus, for example, a 2/3:1/3 chance of $100 or $20 and a 1/2:1/2 chance of $100 or $30 both stochastically dominate a 1/2:1/2 chance of $100 or $20.

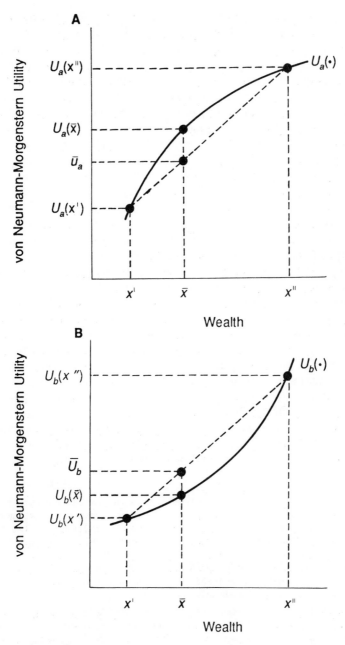

FIGURE 1 Utility functions of risk. A: Concave utility function of a risk averter. B: Convex utility function of a risk lover.

$(1/3)U_a(x'')$ and $\bar{U}_b = (2/3)U_b(x') + (1/3)U_b(x'')$ give its expected *utility* for the utility functions $U_a(\cdot)$ and $U_b(\cdot)$. For the concave (i.e., bowed upward) utility function $U_a(\cdot)$, we have $U_a(\bar{x}) > \bar{u}_a$, which implies that this individual would prefer a sure gain of $\bar{x}$ [which would yield utility $U_a(\bar{x})$] to the gamble. Because someone with a concave utility function will in fact *always* rather receive the expected value of a gamble than receive the gamble itself, concave utility functions are termed *risk averse*. For the convex (bowed downward) utility function $U_b(\cdot)$, we have $\bar{u}_b > U_b(\bar{x})$. Because this preference for bearing the risk rather than receiving the expected value will also extend to all gambles, $U_b(\cdot)$ is termed *risk-loving*. In their famous article, Friedman and Savage (1948) showed how a utility function that was concave at low-wealth levels and convex at high-wealth levels could explain the behavior of individuals who both incur risk by purchasing lottery tickets as well as avoid risk by purchasing insurance.[7] Algebraically, Arrow (1963, 1974), Pratt (1964) and others have shown that the degree of concavity of a utility function, as measured by the curvature index $-U''(x)/U'(x)$, can lead to predictions of how risk attitudes, and hence behavior, will vary with wealth or across individuals in a variety of situations.[8]

Because a knowledge of $U(\cdot)$ would allow the prediction of preferences (and hence behavior) in any risky situation, experimenters and applied decision analysts are frequently interested in eliciting or *recovering* their subjects' (or clients') von Neumann-Morgenstern utility functions. One means of doing this is the *fractile method*. This approach begins by adopting the normalization $U(0) = 0$ and $U(M) = 1$ for some positive amount M and fixing a "mixture probability" $\bar{p}$—say, $\bar{p} = 1/2$. The next step involves obtaining the individual's certainty equivalent ξ_1 of a gamble yielding a 1/2 chance of M and a 1/2 chance of 0, which will have the property that $U(\xi_1) = 1/2$.[9] Finding the certainty equivalent of a gamble yielding a 1/2 chance of ξ_1 and a 1/2 chance of 0 yields the value ξ_2 satisfying $U(\xi_2) = 1/4$.

---

[7] How risk attitudes actually differ over gains versus losses is itself an unsolved problem: evidence consistent with or contradictory to the Friedman-Savage observation of risk seeking over gains and risk aversion over losses can be found in Williams (1966), Kahneman and Tversky (1979), Fishburn and Kochenberger (1979), Grether and Plott (1979), Hershey and Schoemaker (1980a), Payne, Laughhunn, and Crum (1980, 1981), Hershey, Kunreuther, and Schoemaker (1982), and the references cited in these articles. Finally, Feather (1959) and Slovic (1969a) found evidence that subjects' risk attitudes over gains and losses systematically changed when hypothetical situations were replaced by situations involving real money.

[8] For example, if $U_c(\cdot)$ and $U_d(\cdot)$ satisfy $-U_c''(x)/U_c'(x) \geq -U_d''(x)/U_d'(x)$ for all x [i.e., if $U_c(\cdot)$ is at least as risk averse as $U_d(\cdot)$], an individual with utility function $U_c(\cdot)$ would always be willing to pay at least as much as an individual with utility function $U_d(\cdot)$ for (complete) insurance against any risk. See also the related analyses of Ross (1981) and Kihlstrom, Romer, and Williams (1981).

[9] Because the utility of ξ_1 will equal the expected utility of the gamble, it follows that $U(\xi_1) = (1/2)U(M) + (1/2)U(0)$, which under the normalization $U(0) = 0$ and $U(M) = 1$ will equal $1/2$.

Finding the certainty equivalent of a gamble yielding a 1/2 chance of M and a 1/2 chance of ξ_1 yields the value ξ_3 satisfying $U(\xi_3) = 3/4$.[10] By repeating this procedure (i.e., 1/8, 3/8, 5/8, 7/8, 1/16, 3/16, etc.), the utility function can (in the limit) be completely assessed.

To see how the expected utility model can be applied to risk policy, let us consider a disastrous event that is expected to occur with probability p and involve a loss L (L can be measured in either dollars or lives). In many cases, there will be some scope for influencing the magnitudes of either p or L, often at the expense of the other. For example, replacing one large planned nuclear power plant with two smaller, geographically separated plants may (to a first approximation) double the possibility that a nuclear accident will occur. However, the same action may lower the magnitude of the loss (however it is measured) if an accident occurs.

The key tool used in evaluating whether such adjustments should be undertaken is the individual's (or society's) *marginal rate of substitution* $MRS_{p,L}$, which specifies the rate at which an individual (or society) would be willing to trade off a (small) change in p against an offsetting change in L. If the potential adjustment involves better terms than this minimum acceptable rate, it will obviously be preferred; if it involves worse terms, it will not be preferred. Although the exact value of this marginal rate of substitution will depend upon the individual's (or society's) utility function $U(\cdot)$, the expected utility model does offer some general guidance regardless of the shape of the utility function: namely, for a given loss magnitude L, a doubling (tripling, halving, etc.) of the loss probability p should double (triple, half, etc.) the rate at which one would be willing to trade reductions in p against increases in L.[11]

The discussion so far has paralleled the economic literature of the 1960s and 1970s by emphasizing the flexibility of the expected utility model in comparison with the Pascal-Fermat expected value approach. The need to analyze and respond to growing empirical challenges, however, has led economists in the 1980s to concentrate on the behavioral restrictions implied by the expected utility hypothesis. These restrictions are the subject of the next section.

---

[10] As in the previous note, $U(\xi_2) = (1/2)U(\xi_1) + (1/2)U(0)$ and $U(\xi_3) = (1/2)U(M) + (1/2)U(\xi_1)$, which from the normalization $U(0) = 0$, $U(M) = 1$ and the fact that $U(\xi_1) = 1/2$ will equal $1/4$ and $3/4$, respectively.

[11] Because expected utility in this example is given by $\bar{u} = (1 - p)U(W) + pU(W - L)$ (where W is initial wealth or lives), an application of the standard economic formula for the marginal rate of substitution (e.g., see Henderson and Quandt [1980:10–11]) yields $MRS_{p,L} = -(\partial\bar{u}/\partial L)/(\partial\bar{u}/\partial p) = -pU'(W - L)/[U(W) - U(W - L)]$ which, for fixed L, varies proportionately with p.

A Modern Perspective:
Linearity in the Probabilities as a Testable Hypothesis

As a theory of individual behavior, the expected utility model shares many of the underlying assumptions of standard economic consumer theory. In each case, it is assumed that the objects of choice, either commodity bundles or lotteries, can be unambiguously and objectively described and that situations that ultimately imply the same set of availabilities (e.g., the same budget set) will lead to the same choice. In each case, it is also assumed that the individual is able to perform the mathematical operations necessary to actually determine the set of availabilities—for example, to add up the quantities in different sized containers or to calculate the probabilities of compound or conditional events. Finally, in each case, it is assumed that preferences are *transitive*, so that if an individual prefers one object (either a commodity bundle or a risky prospect) to a second, and prefers this second object to a third, he or she will prefer the first object to the third. The validity of these assumptions for choice under uncertainty is examined in later sections.

The strongest and most specific implication of the expected utility hypothesis stems from the form of the expected utility maximand or *preference function* $U(x_1)p_1 + \ldots + U(x_n)p_n$. Although this preference function generalizes the expected value form $x_1p_1 + \ldots + x_np_n$ by dropping the property of linearity in the payoff levels (i.e., the x_i's), it retains the other key property of this form, namely, *linearity in the probabilities*.

Graphically, the property of linearity in the probabilities may be illustrated by considering the set of all lotteries or prospects over some set of fixed outcome levels $x_1 < x_2 < x_3$, which can be represented by the set of all probability triples of the form $P = (p_1, p_2, p_3)$ where $p_i = \text{prob}(x_i)$ and $p_1 + p_2 + p_3 = 1$.[12] Making the substitution $p_2 = 1 - p_1 - p_3$, this set of lotteries can be represented by the points in the unit triangle in the (p_1, p_3) plane, as in Figure 2.[13] Because upward movements in the triangle increase p_3 at the expense of p_2 (i.e., shift probability from the outcome x_2 up to x_3) and leftward movements reduce p_1 to the benefit of p_2 (i.e., shift probability from x_1 up to x_2), these movements (and, more generally, all northwest movements) lead to stochastically dominating lotteries and would accordingly be preferred. For the purposes of illustrating many of the following discussions it will be useful to plot the individual's *indifference curves* in this diagram; that is, the curves in the diagram that

---

[12]Thus, if $x_1 = \$20$, $x_2 = \$30$, and $x_3 = \$100$, the three prospects in footnote 6 would be represented by the points $(p_1, p_3) = (1/3, 2/3)$, $(p_1, p_3) = (0, 1/2)$ and $(p_1, p_3) = (1/2, 1/2)$, respectively.

[13]Although it is fair to describe the renewal of interest in this approach as "modern," modified versions of this triangle diagram can be found as far back as Marschak (1950) and Markowitz (1959:Chap 11).

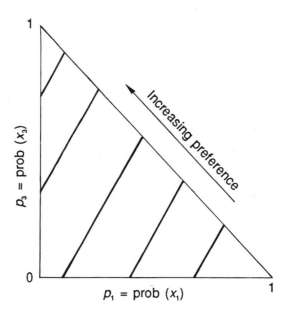

FIGURE 2   Expected utility indifference curves in the triangle diagram.

connect points of equal expected utility.[14] Because each such curve will consist of the set of all (p_1, p_3) points that solve an equation of the form $\bar{u} = U(x_1)p_1 + U(x_2)(1 - p_1 - p_3) + U(x_3)p_3 = k$ for some constant k, and because the probabilities p_1 and p_3 enter linearly (i.e., as multiplicative co-efficients) into this equation, the indifference curves will consist of parallel straight lines, with more preferred indifference curves lying to the northwest. This means that, to know an expected utility maximizer's preferences over the entire triangle, it suffices to know the slope of a single indifference curve.

To see how this diagram can be used to illustrate attitudes toward risk, let us consider Figures 3a and 3b. The dashed lines in the figures are not indifference curves but rather *iso-expected value lines*; that is, lines connecting points with the same expected value that are hence given by the solutions to equations of the form $\bar{x} = x_1p_1 + x_2(1 - p_1 - p_3) + x_3p_3 = k$ for some constant k. Because northeast movements along these lines do not change the expected value of the prospect but do increase the probabilities

---

[14] A useful analogy to the concept of indifference curves is the "constant-altitude" curves on a topographic map, each of which connect points of the same altitude. Just as these curves can be used to determine whether a given movement on the map will lead to a greater or lower altitude, indifference curves can be used to determine whether a given movement in the triangle will lead to greater or lower expected utility.

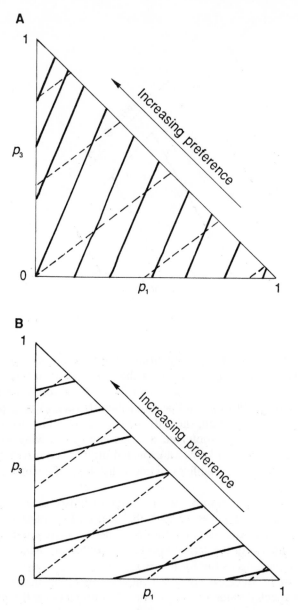

FIGURE 3 A: Relatively steep indifference curves of a risk averter. B: Relatively flat indifference curves of a risk lover.

of the extreme outcomes x_1 and x_3 at the expense of the middle outcome x_2, they are simple examples of *mean preserving spreads* or "pure" increases in risk.[15] When the utility function $U(\cdot)$ is concave (i.e., risk averse), its indifference curves can be shown to be steeper than the iso-expected value lines (Figure 3a),[16] and such increases in risk will lead to less preferred indifference curves. When $U(\cdot)$ is convex (risk loving), its indifference curves will be flatter than the iso-expected value lines (Figure 3b), and these increases in risk will lead to more preferred indifference curves. Finally, if one compares two different utility functions, the one that is more risk averse (in the above Arrow-Pratt sense) will possess the steeper indifference curves.[17]

Behaviorally, the property of linearity in the probabilities can be viewed as a restriction on the individual's preferences over probability mixtures of lotteries. If $\mathbf{P^*} = (p_1^*, ..., p_n^*)$ and $\mathbf{P} = (p_1, ..., p_n)$ are two lotteries over a common outcome set $\{x_1, ..., x_n\}$, the $\alpha : (1-\alpha)$ probability mixture of $\mathbf{P^*}$ and $\mathbf{P}$ is the lottery $\alpha\mathbf{P^*} + (1-\alpha)\mathbf{P} = (\alpha p_1^* + (1-\alpha)p_1, \ldots, \alpha p_n^* + (1-\alpha)p_n)$. This may be thought of as that prospect that yields the same ultimate probabilities over $\{x_1, \ldots, x_n\}$ as the two-stage lottery that offers an $\alpha : (1 - \alpha)$ chance of winning $\mathbf{P^*}$ or $\mathbf{P}$, respectively. It can be shown that expected utility maximizers will exhibit the following property, known as the *independence axiom*:[18]

> If the lottery $\mathbf{P^*}$ is preferred (respectively indifferent) to the lottery $\mathbf{P}$, then the mixture $\alpha\mathbf{P^*} + (1-\alpha)\mathbf{P^{**}}$ will be preferred (respectively indifferent) to the mixture $\alpha\mathbf{P} + (1-\alpha)\mathbf{P^{**}}$ for all $\alpha > 0$ and $\mathbf{P^{**}}$.

This property, which is in fact equivalent to linearity in the probabilities, can be interpreted as follows:

> In terms of the ultimate probabilities over the outcomes $\{x_1, \ldots, x_n\}$, choosing between the mixtures $\alpha\mathbf{P^*} + (1-\alpha)\mathbf{P^{**}}$ and $\alpha\mathbf{P} + (1-\alpha)\mathbf{P^{**}}$ is the same as being offered a coin with a probability $1 - \alpha$ of landing tails, in which case you will obtain the lottery $\mathbf{P^{**}}$, and being asked *before the flip* whether you would rather have $\mathbf{P^*}$ or $\mathbf{P}$ in the event of a head. Now either the coin will land tails, in which case your choice won't have mattered, or else it will land heads, in which case your are "in effect"

---

[15] See, for example, Rothschild and Stiglitz (1970, 1971).

[16] This follows because the slope of the indifference curves can be calculated to be $[U(x_2) - U(x_1)]/[U(x_3) - U(x_2)]$, the slope of the iso-expected value lines can be calculated to be $[x_2 - x_1]/[x_3 - x_2]$, and a concave shape for $U(\cdot)$ implies $[U(x_2) - U(x_1)]/[x_2 - x_1] > [U(x_3) - U(x_2)]/[x_3 - x_2]$ whenever $x_1 < x_2 < x_3$.

[17] Setting his v, w, x, and y equal to x_1, x_2, x_2, and x_3, respectively, this follows directly from theorem 1 of Pratt (1964).

[18] See, for example, Marschak (1950) and Samuelson (1952).

back to a choice between **P*** or **P**, and it is only "rational" to make the same choice as you would before.

Although this is a *prescriptive* argument, it has played a key role in economists' adoption of expected utility as a *descriptive* theory of choice under uncertainty. The mounting evidence against the model has led to a growing tension between those who view economic analysis as the description and prediction of what they consider to be rational behavior and those who view it as the description and prediction of observed behavior. Let us turn now to this evidence.

VIOLATIONS OF LINEARITY IN THE PROBABILITIES

The Allais Paradox and "Fanning Out"

One of the earliest and best-known examples of systematic violation of linearity in the probabilities (or, equivalently, of the independence axiom) is the well-known Allais paradox.[19] This problem involves obtaining the individual's preferred option from each of the following two pairs of gambles (readers who have never seen this problem may want to circle their own choices before proceeding):

$$a_1: \{1.00 \text{ chance of } \$1{,}000{,}000 \quad \text{versus} \quad a_2: \begin{cases} .10 \text{ chance of } \$5{,}000{,}000 \\ .89 \text{ chance of } \$1{,}000{,}000 \\ .01 \text{ chance of } \$0 \end{cases}$$

and

$$a_3: \begin{cases} .10 \text{ chance of } \$5{,}000{,}000 \\ .90 \text{ chance of } \$0 \end{cases} \quad \text{versus} \quad a_4: \begin{cases} .11 \text{ chance of } \$1{,}000{,}000 \\ .89 \text{ chance of } \$0 \end{cases}$$

Defining $\{x_1, x_2, x_3\} = \{\$0; \$1 \text{ million}; \$5 \text{ million}\}$, these four gambles are seen to form to a parallelogram in the (p_1, p_3) triangle (Figures 4a and 4b). Under the expected utility hypothesis, therefore, a preference for a_1 in the first pair would indicate that the individual's indifference curves were relatively steep (as in Figure 4a), which would imply a preference for a_4 in the second pair. In the alternative case of relatively flat indifference curves, the gambles a_2 and a_3 would be preferred.[20] Yet, such researchers as Allais (1953, 1979a), Morrison (1967), Raiffa (1968), and Slovic and Tversky (1974) have found that the most common choice has been for a_1 in the first pair and a_3 in the second, which implies that indifference curves are not parallel but rather *fan out*, as in Figure 4b.

---

[19] See, for example, Allais (1952, 1953, 1979a).

[20] Algebraically, these two cases are equivalent to the expression $[.10 \cdot U(5{,}000{,}000) - .11 \cdot U(1{,}000{,}000) + .01 \cdot U(0)]$, being respectively negative or positive.

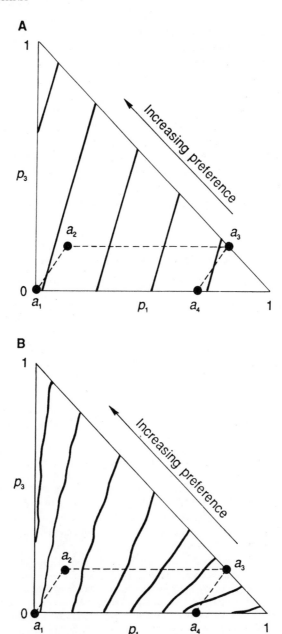

FIGURE 4   A: Expected utility indifference curves and the Allais Paradox. B: Indifference curves that "fan out" and the Allais Paradox.

One of the criticisms of this evidence has been that individuals whose choices violated the independence axiom would "correct" themselves once the nature of their violations were revealed by an application of the above coin-flip argument.[21] Thus, although even Savage chose a_1 and a_3 when he was first presented with this problem, upon reflection, he concluded that these preferences were in error.[22] Although his own reaction was undoubtedly sincere, the prediction that individuals would *invariably* react in such a manner has not been sustained in direct empirical testing. In experiments in which subjects were asked to respond to Allais-type problems and then presented with written arguments both for and against the expected utility position, neither MacCrimmon (1968), Moskowitz (1974), nor Slovic and Tversky (1974) found predominant net swings toward the expected utility choices.[23]

Additional Evidence of Fanning Out

Although the Allais paradox was originally dismissed as an isolated example, it is now known to be a special case of a general empirical pattern that is called the *common consequence* effect. This effect involves pairs of probability mixtures of the form

$$b_1: \left\{ \begin{array}{l} \alpha \;\; \text{chance of x} \\ 1-\alpha \; \text{chance of P**} \end{array} \right. \quad \text{versus} \quad b_2: \left\{ \begin{array}{l} \alpha \;\; \text{chance of P} \\ 1-\alpha \; \text{chance of P**} \end{array} \right.$$

and

$$b_3: \left\{ \begin{array}{l} \alpha \;\; \text{chance of x} \\ 1-\alpha \; \text{chance of P*} \end{array} \right. \quad \text{versus} \quad b_4: \left\{ \begin{array}{l} \alpha \;\; \text{chance of P} \\ 1-\alpha \; \text{chance of P*} \end{array} \right.$$

---

[21] Let **P** be a gain of $1 million, let **P*** be a (10/11):(1/11) chance of $5 million or $0, and let $\alpha = .11$. The choice between a_1 and a_2 is then equivalent to a choice between $\alpha P + (1 - \alpha)P**$ and $\alpha P* + (1 - \alpha)P**$ when **P**** is a gain of $1 million; the choice between a_4 and a_3 is a choice between $\alpha P + (1 - \alpha)P**$ and $\alpha P* + (1 - \alpha)P**$ when **P**** is a gain of $0. Thus one should choose a_1 and a_4 if **P** is preferred to **P*** or a_2 and a_3 if **P*** is preferred to **P**.

[22] Reports of this incident can be found in Savage (1954:101-103) and Allais (1979b:533-535). In that instance the payoffs of {$0;$1 million;$5 million} in the displayed example were replaced by {$0;$500,000;$2.5 million} (1952 dollars).

[23] In each of MacCrimmon's experiments, for example, he obtained approximately 60 percent conformity with the independence axiom (1968:7-11). However, when presented with opposing written arguments, the pro-expected utility argument was chosen by only 20 percent of the subjects in the first experiment and 50 percent of the subjects in the second experiment (subjects in the third experiment were not presented with written arguments). In subsequent interviews with the experimenter, the percentage of subjects conforming to the independence axiom did rise to 75 percent. Although MacCrimmon did not apply pressure to induce the subjects to adopt expected utility and "repeatedly emphasized that there was no right or wrong answer," he personally believed in "the desirability of using the [expected utility] postulates in training decision makers" (1968:21-22), a fact that Slovic and Tversky felt "may have influenced the subjects to conform to the axioms" (1974:369).

where **P** involves outcomes both greater and less than x, and **P**** stochastically dominates **P***.[24] Although the independence axiom clearly implies choices of either b_1 and b_3 (if x is preferred to **P**) or b_2 and b_4 (if **P** is preferred to x), researchers have again found a tendency for subjects to choose b_1 in the first pair and b_4 in the second.[25] When the distributions **P**, **P***, and **P**** are each over a common outcome set $\{x_1, x_2, x_3\}$ that includes x, the prospects b_1, b_2, b_3, and b_4 will again form a parallelogram in the (p_1, p_3) triangle, and a choice of b_1 and b_4 again implies indifference curves that fan out as in Figure 4b.

The intuition behind this phenomenon can be described in terms of the coin-flip scenario noted earlier. According to the independence axiom, preferences over what would occur in the event of heads should not depend upon what would occur in the event of tails. In fact, however, they may well depend on what would otherwise happen.[26] The common consequence effect states that the better off individuals would be in the event of tails (in the sense of stochastic dominance), the more risk averse they become over what they would receive in the event of heads. Intuitively, if the distribution **P**** in the pair $\{b_1, b_2\}$ involves very high outcomes, an individual may prefer not to bear further risk in the unlucky event that he or she does not receive it, and prefer instead the sure outcome x over the distribution **P** in this event (i.e., choose b_1 over b_2). If **P*** in $\{b_3, b_4\}$ involves very low outcomes, however, an individual may be more willing to bear risk in the (lucky) event that he or she doesn't receive it, and prefer the lottery **P** to the outcome x in this case (i.e., choose b_4 over b_3). Note that it is not the individual's beliefs regarding the probabilities in **P** that are affected here, merely his or her willingness to bear them.[27]

A second class of systematic violations, stemming from another early example of Allais (1953), is known as the *common ratio effect*. This phenomenon involves pairs of prospects of the form

$$c_1: \left\{ \begin{array}{l} p \text{ chance of } \$X \\ 1-p \text{ chance of } \$0 \end{array} \right. \quad \text{versus} \quad c_2: \left\{ \begin{array}{l} q \text{ chance of } \$Y \\ 1-q \text{ chance of } \$0 \end{array} \right.$$

[24]The Allais Paradox choices a_1, a_2, a_3, and a_4 correspond to b_1, b_2, b_4, and b_3, where $\alpha = .11$, $x = \$1$ million, **P** is a $(10/11){:}(1/11)$ chance of \$5 million or \$0, **P*** $=$ is a sure gain of \$0, and **P**** is a sure gain of \$1 million.

[25]See MacCrimmon (1968), MacCrimmon and Larsson (1979), Kahneman and Tversky (1979) and Chew and Waller (1986).

[26]As Bell (1985) notes, "winning the top prize of \$10,000 in a lottery may leave one much happier than receiving \$10,000 as the lowest prize in a lottery."

[27]In a conversation with the author, Kenneth Arrow has offered an alternative phrasing of this argument: The widely maintained hypothesis of decreasing absolute risk aversion asserts that individuals will display more risk aversion in the event of a loss and less risk aversion in the event of a gain. In the common consequence effect, individuals display more risk aversion in the event of an *opportunity* loss, and less risk aversion in the event of an *opportunity* gain.

and

$$c_3: \begin{cases} \alpha p & \text{chance of } \$X \\ 1-\alpha p & \text{chance of } \$0 \end{cases} \quad \text{versus} \quad c_4: \begin{cases} \alpha q & \text{chance of } \$Y \\ 1-\alpha q & \text{chance of } \$0 \end{cases}$$

where $p > q, 0 < X < Y$ and $0 < \alpha < 1$; it includes the "certainty effect" of Kahneman and Tversky (1979) and the ingenious "Bergen paradox" of Hagen (1979) as special cases.[28] Setting $\{x_1, x_2, x_3\} = \{0, X, Y\}$ and plotting these prospects in the (p_1, p_3) triangle, the segments $\overline{c_1 c_2}$ and $\overline{c_3 c_4}$ are seen to be parallel (as in Figure 5a), so that the expected utility model again predicts choices of c_1 and c_3 (if the individual's indifference curves are steep) or c_2 and c_4 (if they are flat). Yet, experimental studies have found a systematic tendency for choices to depart from these predictions in the direction of preferring c_1 and c_4,[29] which again suggests that indifference curves fan out, as in the figure. In a variation on this approach, Kahneman and Tversky (1979) replaced the gains of $\$X$ and $\$Y$ in the above gambles with losses of these magnitudes and found a tendency to depart from expected utility in the direction of c_2 and c_3. Defining $\{x_1, x_2, x_3\}$ as $\{-Y, -X, 0\}$ (to maintain the ordering $x_1 < x_2 < x_3$) and plotting these gambles in Figure 5b, a choice of c_2 and c_3 is again seen to imply that indifference curves fan out. Finally, Battalio, Kagel, and MacDonald (1985) found that laboratory rats choosing among gambles that involved substantial variations in their actual daily food intake also exhibited this pattern of choices.

A third class of evidence stems from the elicitation method described in the previous section. In particular, the reader should note that there is no reason why the mixture probability $\bar{p}$ *must* be 1/2, as in the earlier example. Picking any other value—say $\bar{p}* = 1/4$—and obtaining the individual's certainty equivalent ξ_1^* of the gamble offering a 1/4 chance of M and a 3/4 chance of 0 will lead to the property that $U(\xi_1^*) = 1/4$; in addition, just as in the previous case of $\bar{p} = 1/2$, the procedure using $\bar{p}* = 1/4$ (or any other fixed value) can also be continued to (in the limit) completely recover $U(\cdot)$.

Although this procedure should recover the same (normalized) utility function for any value of the mixture probability $\bar{p}$, such researchers as Karmarkar (1974, 1978) and McCord and de Neufville (1983, 1984) have found a tendency for higher values of $\bar{p}$ to lead to the "recovery" of higher valued utility functions (Figure 6a). By illustrating the gambles used to

---

[28] The former involves setting $p = 1$, and the latter consists of a two-step choice problem in which individuals exhibit the effect with $Y = 2X$ and $p = 2q$. Kahneman and Tversky (1979), for example, found that 80 percent of their subjects preferred a sure gain of 3,000 Israeli pounds to a .80 chance of winning 4,000; 65 percent, however, preferred a .20 chance of winning 4,000 to a .25 chance of winning 3,000. The name "common ratio effect" comes from the common value of prob(X)/prob(Y) in the pairs $\{c_1, c_2\}$ and $\{c_3, c_4\}$.

[29] See Tversky (1975), MacCrimmon and Larsson (1979), and Chew and Waller (1986).

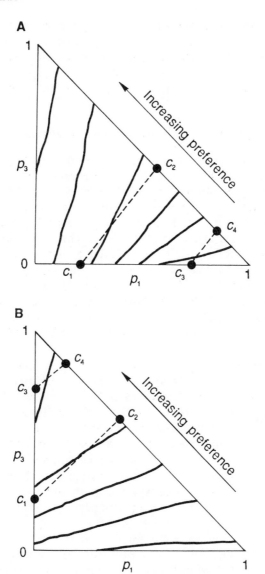

FIGURE 5 A: Indifference curves that fan out and the common ratio effect. B: Indifference curves that fan out and the common ratio effect with negative payoffs.

obtain the certainty equivalents ξ_1, ξ_2, and ξ_3 for the mixture probability $\bar{p} = 1/2$, ξ_1^* for $\bar{p}* = 1/4$, and ξ_1^{**} for $\bar{p}** = 3/4$, Figure 6b shows that, as with the common consequence and common ratio effects, this utility evaluation effect is precisely what would be expected from an individual whose indifference curves departed from expected utility by fanning out.[30]

Non-Expected Utility Models of Preferences

The systematic nature of these departures from linearity in the probabilities have led several researchers to generalize the expected utility model by positing nonlinear functional forms for the individual preference function. Some examples of such forms and researchers who have studied them are given in Table 1. Many (though not all) of these forms are flexible enough to exhibit the properties of stochastic dominance preference, risk aversion/risk preference, and fanning out, and the Chew/MacCrimmon/Fishburn and Quiggin forms have proven to be particularly useful both theoretically and empirically. Additional analyses of the above forms can be found in Chew, Karni, and Safra (1987); Fishburn (1982, 1984a,b); Röell (1987); Segal (1984, 1987); and Yaari (1987). For general surveys of these models, see Machina (1983a), Sugden (1986), and Weber and Camerer (1987).

Although such forms allow for the modeling of preferences that are more general than those allowed by the expected utility hypothesis, each requires a different set of conditions on its component functions $\nu(\cdot)$, $\pi(\cdot)$, $\tau(\cdot)$ or $g(\cdot)$ for the properties of stochastic dominance preference, risk aversion/risk preference, comparative risk aversion, and so forth. In particular, the standard expected utility results that link properties of the function $U(\cdot)$ to such aspects of behavior generally will *not* extend to the corresponding properties of the function $\nu(\cdot)$ in the above forms. Does this imply that the study of non-expected utility preferences requires one to abandon the vast body of theoretical results and intuition that have been developed within the expected utility framework?

Fortunately, the answer is no. An alternative approach to the analysis of non-expected utility preferences proceeds not by adopting a *specific* nonlinear function but by considering nonlinear functions *in general*, and using calculus to extend results from expected utility theory in the same manner in which it is typically used to extend results involving linear

---

[30]Having found the value ξ_1 that solves $U(\xi_1) = (1/2)U(M) + (1/2)U(0)$, let us now choose $\{x_1, x_2, x_3\} = \{0, \xi_1, M\}$, so that the indifference curve through the point $(0,0)$ (i.e., a sure gain of ξ_1) also passes through the point $(1/2, 1/2)$ (a 1/2:1/2 chance of M or 0). The ordering of the values $\xi_1, \xi_2, \xi_3, \xi_1^*$ and ξ^**_1 in Figure 6a is derived from the individual's preference ordering over the five distributions in Figure 6b for which they are the respective certainty equivalents.

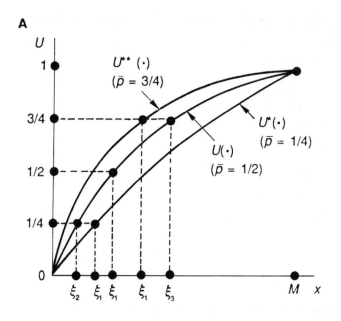

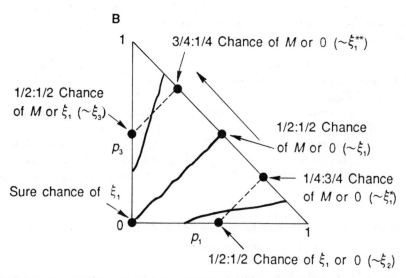

FIGURE 6   A: "Recovered" utility functions for mixture probabilities 1/4, 1/2, and 3/4. B: Fanning out indifference curves that generate the responses of Figure 6a. Note: ~ denotes indifference.

TABLE 1 Examples of Non-Expected Utility Preference Functions

Function	Researcher
$\nu(x_1)\pi(p_1) + \cdots + \nu(x_n)\pi(p_n)$	Edwards (1955, 1962); Kahneman and Tversky (1979)
$\dfrac{\nu(x_1)\pi(p_1) + \cdots + \nu(x_n)\pi(p_n)}{\pi(p_1) + \cdots + \pi(p_n)}$	Karmarkar (1978, 1979)
$\dfrac{\nu(x_1)p_1 + \cdots + \nu(x_n)p_n}{\tau(x_1)p_1 + \cdots + \tau(x_n)p_n}$	Chew and MacCrimmon (1979a,b); Chew (1983); Fishburn (1983)
$\nu(x_1)g(p_1) + \nu(x_2)[g(p_2 + p_1) - g(p_1)] + \nu(x_3)[g(p_3 + p_2 + p_1) - g(p_2 + p_1)] + \cdots$	Quiggin (1982)
$[\nu(x_1)p_1 + \cdots + \nu(x_n)p_n] + [\pi(x_i)p_1 + \cdots + \pi(x_n)p_n]^2$	Machina (1982)

functions. (Readers who are not interested in the details of this approach may wish to skip ahead to the next section.[31])

Specifically, let us consider the set of all probability distributions $\mathbf{P}$ = $(p_1, \ldots, p_n)$ over a fixed outcome set $\{x_1, \ldots, x_n\}$, so that the expected utility preference function can be written as $V(\mathbf{P}) = V(p_1, \ldots, p_n) = U(x_1)p_1 + \ldots + U(x_n)p_n$. Let us also think of $U(x_i)$ not as a "utility level" but rather as the coefficient of $p_i = \text{prob}(x_i)$ in this linear function. If these coefficients are plotted against x_i as in Figure 7, the expected utility results of the previous section can be stated as:

- *Stochastic Dominance Preference:* $V(\cdot)$ will exhibit global stochastic dominance preference if and only if the coefficients $\{U(x_i)\}$ are increasing in x_i, as in Figure 7.
- *Risk Aversion:* $V(\cdot)$ will exhibit global risk aversion if and only if the coefficients $\{U(x_i)\}$ are concave in x_i,[32] as in Figure 7.
- *Comparative Risk Aversion:* The expected utility preference function $V^*(\mathbf{P}) = U^*(x_1)p_1 + \ldots + U^*(x_n)p_n$ will be at least as risk averse as $V(\cdot)$ if and only if the coefficients $\{U^*(x_i)\}$ are at least as concave in x_i as $\{U(x_i)\}$.[33]

---

[31] More complete developments of this approach may be found in Machina (1982, 1983b).

[32] As in footnote 16, this is equivalent to the condition that $[U(x_{i+1}) - U(x_i)]/[x_{i+1} - x_i] < [U(x_i) - U(x_{i-1})]/[x_i - x_{i-1}]$ for all i.

[33] This is equivalent to the condition that $U^*(x_i) \equiv \rho(U(x_i))$ for some increasing concave function $\rho(\cdot)$.

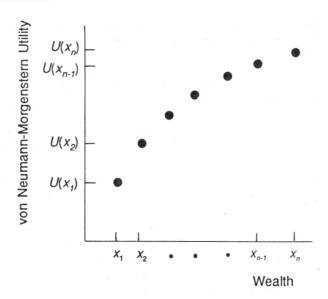

FIGURE 7 von Neumann-Morgenstern utilities as coefficients of the expected utility preference function $V(p_1,...,p_n) = U(x_1)p_1 + ... + U(x_n)p_n$.

Now, let us consider the case in which the individual's preference function $\mathcal{V}(\mathbf{P}) = \mathcal{V}(p_1,\ldots,p_n)$ is not linear (i.e., not expected utility) but at least differentiable, and let us consider its partial derivatives $\mathcal{U}(x_i;\mathbf{P})$ $= \partial \mathcal{V}(\mathbf{P})/\partial p_i = \partial \mathcal{V}(\mathbf{P})/\partial \text{prob}(x_i)$. Some probability distribution $\mathbf{P}_0$ can be chosen and these $\mathcal{U}(x_i;\mathbf{P}_0)$ values plotted against x_i. If they are increasing in x_i, it is clear that any *infinitesimal* stochastically dominating shift from $\mathbf{P}_0$, such as a decrease in some p_i and matching increase in p_{i+1}, will be preferred. If they are concave in x_i, any *infinitesimal* mean preserving spread, such as a drop in p_i and (mean preserving) rise in p_{i-1} and p_{i+1}, will make the individual worse off. In light of this correspondence between the coefficients $\{U(x_i)\}$ of the expected utility preference function $V(\cdot)$ and the partial derivatives $\{\mathcal{U}(x_i;\mathbf{P}_0)\}$ of the non-expected utility preference function $\mathcal{V}(\cdot)$, $\{\mathcal{U}(x_i;\mathbf{P}_0)\}$ as the individual's local utility indices at $\mathbf{P}_0$.

Of course, the above results will only hold exactly for infinitesimal shifts from the distribution $\mathbf{P}_0$. However, another result from standard calculus can be exploited to show how "expected utility" results may be applied to the exact global analysis of non-expected utility preferences. The reader should recall that, in many cases, a differentiable function will exhibit a specific global property if and only if that property is exhibited by its linear approximations at each point. For example, a differentiable function will be globally nondecreasing if and only if its linear approximation at each point is nonnegative. In fact, most of the fundamental properties of risk attitudes and their expected utility characterizations are precisely of this type. In particular, the following can be shown:

- *Stochastic Dominance Preference:* A non-expected utility preference function $V(\cdot)$ will exhibit global stochastic dominance preference if and only if its local utility indices $\{U(x_i;P)\}$ are increasing in x_i at each distribution **P**.
- *Risk Aversion:* $V(\cdot)$ will exhibit global risk aversion if and only if its local utility indices $\{U(x_i;P)\}$ are concave in x_i at each distribution **P**.
- *Comparative Risk Aversion:* The preference function $V^*(\cdot)$ will be globally at least as risk averse[34] as $V(\cdot)$ if and only if its local utility indices $\{U^*(x_i;P)\}$ are at least as concave in x_i as $\{U(x_i:P)\}$ at each **P**.

Figures 8a and 8b are a graphic illustration of this approach for the outcome set $\{x_1, x_2, x_3\}$. Here, the solid curves denote the indifference curves of the non-expected utility preference function $V(P)$. The parallel lines near the lottery $\mathbf{P_0}$ denote the tangent "expected utility" indifference curves that correspond to the local utility indices $\{U(x_i;P_0)\}$ at $\mathbf{P_0}$. As always with differentiable functions, an infinitesimal change in the probabilities at $\mathbf{P_0}$ will be preferred if and only if it would be preferred by this tangent linear (i.e., expected utility) approximation. Figure 8b illustrates the above "risk aversion" result. It is clear that these indifference curves will be globally risk averse (averse to mean preserving spreads) if and only if these are everywhere steeper than the dashed iso-expected value lines. However, this is equivalent to all of their tangents being steeper than these lines, which in turn is equivalent to all of their local expected utility approximations being steeper—or, in other words, to the local utility indices $\{U(x_i;P)\}$ being concave in x_i at each distribution **P**.

My fellow researchers and I have shown how this and similar techniques can be applied to further extend the results of expected utility theory to the case of non-expected utility preferences, to characterize and explore the implications of preferences that "fan out," and to conduct new and more general analyses of economic behavior under uncertainty.[35] Still, although I feel that they constitute a useful and promising response to the phenomenon of nonlinearities in the probabilities, these models do *not* provide solutions to the more problematic empirical phenomena described in the following sections.

---

[34]For the appropriate generalizations of the expected utility concepts of "at least as risk averse" in this result, see Machina (1982, 1984).

[35]See, for example, Machina (1982, 1983b, 1984); Chew (1983); Fishburn (1984a); Epstein (1985); Dekel (1986); Allen (1987); Chew, Karni, and Safra (1987); Karni and Safra (1987), and Machina and Neilson (1987).

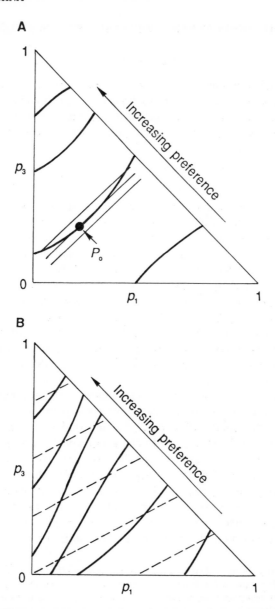

FIGURE 8   A: Tangent "expected utility" approximation to non-expected utility indifference curves. Note: Solid lines are local expected utility approximation to non-expected utility indifference curves at P_0. B: Risk aversion of every local expected utility approximation is equivalent to global risk aversion. Note: Dashed lines are iso-expected value lines.

THE PREFERENCE REVERSAL PHENOMENON

The Evidence

The finding now known as the preference reversal phenomenon was initially reported by psychologists Lichtenstein and Slovic (1971). In this study, subjects were first presented with a number of pairs of bets and asked to choose one bet out of each pair. Each of these pairs took the following form:

$$\text{P-bet:}\left[\begin{array}{l} p \quad \text{chance of \$X} \\ 1\text{-}p \text{ chance of \$x} \end{array}\right. \quad \text{versus} \quad \text{\$-bet:}\left[\begin{array}{l} q \quad \text{chance of \$Y} \\ 1\text{-}q \text{ chance of \$y} \end{array}\right.$$

where X and Y are respectively greater than x and y, p is greater than q, and Y is greater than X (the names "P-bet" and "$-bet" come from the greater probability of winning in the first bet and greater possible gain in the second). In some cases, x and y took on small negative values. The subjects were next asked to "value" (state certainty equivalents for) each of these bets. The different valuation methods that were used consisted of (a) asking subjects to state their minimum selling price for each bet if they were to own it, (b) asking them to state their maximum bid price for each bet if they were to buy it, and (c) the elicitation procedure of Becker, DeGroot, and Marschak (1964), in which it is in a subject's best interest to reveal his or her true certainty equivalent.[36] In the latter case, real money was in fact used.

The expected utility model, as well as each of the non-expected utility models of the previous section, clearly implies that the bet that is actually chosen out of each pair will also be the one that is assigned the higher certainty equivalent.[37] However, Lichtenstein and Slovic (1971) found a systematic tendency to violate this prediction in the direction of choosing the P-bet in a direct choice but assigning a higher value to the $-bet. In one experiment, for example, 127 out of 173 subjects assigned a higher sell price to the $-bet in *every* pair in which the P-bet was chosen. Similar findings were obtained by Lindman (1971) and, in an interesting variation on the usual experimental setting, by Lichtenstein and Slovic (1973) in a

---

[36] Roughly speaking, the subject states a value for the item and then the experimenter draws a random price. If the price is above the stated value, the subject forgoes the item and receives the price. If the drawn price is below the stated value, the subject keeps the item. The reader can verify that under such a scheme, it will never be in a subject's best interest to report anything other than his or her true value.

[37] Economic theory tells us that income effects may well lead an individual to assign a lower bid price to the object that, if both were free, would actually be preferred. However, such an effect will *not* apply to either selling prices or the Becker, DeGroot, and Marschak procedure. For discussions of the empirical evidence on sell price/bid price disparities, see Knetsch and Sinden (1984) and the references cited there.

Las Vegas casino where customers actually staked (and hence sometimes lost) their own money. In another real-money experiment, Mowen and Gentry (1980) found that groups who could discuss their (joint) decisions were if anything more likely than individuals to exhibit the phenomenon.

Although these above studies involved deliberate variations in design in order to check for the robustness of this phenomenon, they were nevertheless received skeptically by economists, who perhaps not unnaturally felt they had more at stake than psychologists in this type of finding. In an admitted attempt to "discredit" this work, economists Grether and Plott (1979) designed a pair of experiments in which they corrected for issues of incentives, income effects,[38] strategic considerations, the ability to indicate indifference, and so forth. They expected that the experiments would not generate this phenomenon, but they nonetheless found it in both. Further design modifications by Pommerehne, Schneider, and Zweifel (1982) and Reilly (1982) yielded the same results. Finally, the phenomenon has been found to persist (although in mitigated form) even when subjects are allowed to engage in experimental market transactions involving the gambles (Knez and Smith, 1987), or when the experimenter is able to act as an arbitrager and make money from such reversals (Berg, Dickhaut, and O'Brien, 1983).

Two Interpretations of This Phenomenon

How one interprets these findings depends on whether one adopts the world view of an economist or a psychologist. An economist would reason as follows: Each individual possesses a unique underlying preference ordering over objects (in this case lotteries), and information about this preference ordering can be gleaned from either direct choice questions or (properly designed) valuation questions.[39] Someone exhibiting the preference reversal phenomenon is therefore indicating that (a) they are indifferent regarding the choice between the P-bet and some sure amount ξ_P, (b) they strictly prefer the P-bet to the $-bet, and (c) they are indifferent regarding the choice between the $-bet and an amount $\xi_\$$ *greater than* ξ_p. Assuming that they in fact prefer $\xi_\$$ to the lesser amount ξ_p, this implies that their preferences over these four objects are cyclic or intransitive.

---

[38] In addition to the problem with bid prices discussed in the previous note, Grether and Plott (1979) noted that subjects' changing wealth (as a result of the actual play of these gambles during the course of the experiment), or the changing of their *expected* wealth (in those experiments in which chosen gambles would be played at the end), could be a source of income effects.

[39] Formally, this ordering is represented by the individual's weak preference relation $\succeq$, where "$A \succeq B$" is read "A is at least as preferred as B." From this one may in turn derive the individual's *strict preference relation* $\succ$ and *indifference relation* $\sim$, where "$A \succ B$" denotes that $A \succeq B$ but *not* $B \succeq A$, and "$A \sim B$" denotes that both $A \succeq B$ and $B \succeq A$.

Psychologists, on the other hand, would deny the premise of an common underlying mechanism generating both choice and valuation behavior. Rather, they view choice and valuation (even different forms of valuation) as distinct processes, subject possibly to different influences. In other words, individuals exhibit what are termed *response mode effects*. Excellent discussions and empirical examinations of this phenomenon and its implications for the elicitation of both probabilistic beliefs and utility functions can be found in Hogarth (1975, 1980); Hershey, Kunreuther, and Schoemaker (1982); Slovic, Fischhoff, and Lichtenstein (1982); Hershey and Schoemaker (1985); and MacCrimmon and Wehrung (1986). To report how the response mode study of Slovic and Lichtenstein (1968) actually led them to predict the preference reversal phenomenon, I can do no better than to quote the authors themselves:

> The impetus for this study [Lichtenstein and Slovic (1971)] was our observation in our earlier 1968 article that choices among pairs of gambles appeared to be influenced primarily by probabilities of winning and losing, whereas buying and selling prices were primarily determined by the dollar amounts that could be won or lost. . . . In our 1971 article, we argued that, if the information in a gamble is processed differently when making choices and setting prices, it should be possible to construct pairs of gambles such that people would choose one member of the pair but set a higher price on the other. [Slovic and Lichtenstein (1983:597)]

Implications of the Economic World View

The issue of intransitivity is new neither to economics nor to choice under uncertainty. May (1954), for example, observed intransitivities in pairwise rankings of three alternative marriage partners, in which each candidate was rated highly in two of three attributes (intelligence, looks, wealth) and low in the third. In an uncertain context, Blyth (1972) has adapted this approach to construct a set of random variables $(\tilde{x}, \tilde{y}, \tilde{z})$ such that $\text{prob}(\tilde{x} > \tilde{y}) = \text{prob}(\tilde{y} > \tilde{z}) = \text{prob}(\tilde{z} > \tilde{x}) = 2/3$, so that individuals making pairwise choices on the basis of these probabilities would also be intransitive. In addition to the preference reversal phenomenon, Edwards (1954a)[40] and Tversky (1969) have also observed intransitivities in preferences over risky prospects. On the other hand, researchers have also shown that many aspects of economic theory, in particular the existence of demand functions and of general equilibrium, are surprisingly robust to the phenomenon of intransitivity (Sonnenschein, 1971; Mas-Colell (1974); Shafer, 1974, 1976; Kim and Richter, 1986; Epstein, 1987).

---

[40] See also the discussions of these findings by Edwards (1954b:404–405), Davis (1958:28), and Weinstein (1968:337).

In any event, economists have begun to develop and analyze models of nontransitive preferences over lotteries. The leading example of this is the "regret theory" model developed independently by Bell (1982, 1983) (see also Bell and Raiffa [1980]), Fishburn [1981, 1982, 1984a,b], and Loomes and Sugden [1982, 1983a,b]). In this model of pairwise choice the von Neumann-Morgenstern utility function $U(x)$ is replaced by a *regret/rejoice function* $r(x, y)$ that represents the level of satisfaction (or, if negative, dissatisfaction) the individual would experience if he or she were to receive the outcome x when the alternative choice would have yielded the outcome y (this function is assumed to satisfy $r(x, y) = -r(y, x)$ for all values of x and y). In choosing between statistically independent gambles $\mathbf{P} = (p_1, \ldots, p_n)$ and $\mathbf{P^*} = (p_1^*, \ldots, p_n^*)$ over a common outcome set $\{x_1, \cdots, x_n\}$, the individual will choose $\mathbf{P^*}$ if the expected value of the function $r(x, y)$ is positive and $\mathbf{P}$ if it is negative[41] (Once again, readers who wish to skip the mathematical details of this approach may proceed to the following subsection.)

It is interesting to note that when the regret/rejoice function takes the special form $r(x, y) = U(x) - U(y)$ this model reduces to the expected utility model.[42] In general, however, such an individual will neither be an expected utility maximizer nor have transitive preferences.

Yet, this intransitivity does not prevent the graphing of such preferences or even the application of the "expected utility" analysis to them. To see the former, let us consider the case in which the individual is facing alternative independent lotteries over a common outcome set $\{x_1, x_2, x_3\}$, so that the triangle diagram may again be used to illustrate their "indifference curves," which will appear as in Figure 9. In such a case, it is important to understand what is and is not still true of these indifference curves. The curve through $\mathbf{P}$ will still correspond to the points (i.e., lotteries) that are indifferent to $\mathbf{P}$, and it will still divide the points that are strictly preferred to $\mathbf{P}$ (the points in the direction of the arrow) from the ones to which $\mathbf{P}$ is strictly preferred. Furthermore, if (as in the figure) $\mathbf{P^*}$ lies above the indifference curve through $\mathbf{P}$, then $\mathbf{P}$ will lie below the indifference curve through $\mathbf{P^*}$ (i.e., the individual's ranking of $\mathbf{P}$ and $\mathbf{P^*}$ will be unambiguous). Unlike indifference curves for transitive preferences, however, these curves

---

[41]Algebraically, this expected value is given by the double summation $\sum_i \sum_j r(x_i, x_j) p_i^* p_j$.

[42]This follows because $\sum_i \sum_j r(x_i, x_j) p_i^* p_j = \sum_i \sum_j [U(x_i) - U(x_j)] p_i^* p_j = \sum_i U(x_i) p_i^* - \sum_j U(x_j) p_j$, so that the individual will prefer $\mathbf{P^*}$ to $\mathbf{P}$ if and only if $\sum_i U(x_i) p_i^* > \sum_j U(x_j) p_j$. When $r(\cdot, \cdot)$ takes the form $r(x, y) = \nu(x)\tau(y) - \nu(y)\tau(x)$, the expected regret model reduces to the (transitive) Chew/MacCrimmon/Fishburn form of Table 1. This is the most general form of the model compatible with transitivity.

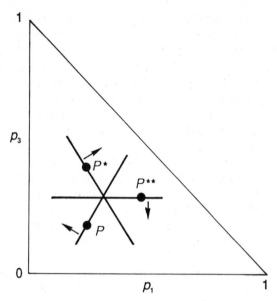

FIGURE 9   "Indifference curves" for the expected regret model.

will cross,[43] and preferences over the lotteries **P**, **P***, and **P**** are seen to form an intransitive cycle. In regions in which the indifference curves do *not* cross (such as near the origin), however, the individual will be indistinguishable from someone with transitive (albeit non-expected utility) preferences.

To see how expected utility results can be extended to this nontransitive framework, let us fix a lottery $P = (p_1, \ldots, p_n)$ and consider the question of when an (independent) lottery $P^* = (p_1^*, \ldots, p_n^*)$ will be preferred or not preferred to P. Defining the "utility function" $\phi(x;P) = r(x, x_1)p_1 + \ldots + r(x, x_n)p_n$, it is possible to show that **P*** will be preferred to **P** if and only if $\phi(x_1;P)p_1^* + \ldots + \phi(x_n;P)p_n^* > \phi(x_1;P)p_1 + \ldots + \phi(x_n;P)p_n$—in other words, if and only if **P*** implies a higher expectation of the function $\phi(x;P)$ than does P.[44] Thus, if $\phi(x;P)$ is increasing in x for all lotteries **P**, the individual will exhibit global stochastic dominance preference; if $\phi(x;P)$ is concave in x for all **P**, the individual will exhibit global risk aversion, even though he or she is not necessarily transitive (these conditions will clearly be satisfied

---

[43] In this model, the indifference curves will necessarily all cross at the same point. This (unique) point will accordingly be ranked indifferent to all lotteries in the triangle.

[44] Because $r(x,y) = -r(y,x)$ for all x and y implies $\sum_i \sum_j r(x_i,x_j)p_i p_j = 0$, **P*** will be preferred to **P** if and only if $0 < \sum_i \sum_j r(x_i,x_j)p_i^* p_j = \sum_i \sum_j r(x_i,x_j)p_i^* p_j - \sum_i \sum_j r(x_i,x_j)p_i p_j$
$= \sum_i [\sum_j r(x_i,x_j)p_j]p_i^* - \sum_i [\sum_j r(x_i,x_j)p_j]p_i = \sum_i \phi(x_i;P)p_i^* - \sum_i \phi(x_i;P)p_i.$

if $r(x, y)$ is increasing and concave in x).[45] The analytics of expected utility theory are robust, indeed.

Bell, Raiffa, Loomes, Sugden, and Fishburn have also shown how specific assumptions about the form of the regret/rejoice function will generate the common consequence effect, the common ratio effect, the preference reversal phenomenon, and other observed properties of choice over lotteries.[46] The theoretical and empirical prospects for this approach seem quite impressive.

Implications of the Psychological World View

On the other hand, how should economists respond if it turns out that the psychologists are right and that the preference reversal phenomenon really *is* generated by some form of response mode effect (or effects)? In that case, the first thing to do would be to try to determine if there were analogues of such effects in real-world economic situations.[47] Will individuals behave differently when they are determining their valuation of an object (e.g., reservation bid on a used car) than they will when reacting to a fixed and nonnegotiable price for the same object? Because a proper test of this question would require correcting for any possible strategic or information-theoretic (e.g., signaling) issues, it would not be a simple undertaking. However, in light of the experimental evidence, I feel it is crucial that it be attempted.

Let us say that it was found that response mode effects did not occur outside of the laboratory. In that case, we scientists could rest more easily, although we could not forget about such issues completely: experimenters testing *other* economic theories and models (e.g., auctions) would have to be forever mindful of the possible influence of the particular response mode used in their experimental design.

On the other hand, what if response mode effects *were* found out in the field? In such circumstances, we would want to determine, perhaps by going back to the laboratory, whether the rest of economic theory remained valid—provided the response mode were held constant. If this

---

[45] It is important to note that although the function $\phi(x_i;\mathbf{P})$ plays a role very similar to the local utility index $\mathcal{U}(x_i;\mathbf{P})$, it is a different concept. Unlike the linear approximation to a nonlinear preference function, the previous inequality is both exact and global.

[46] Loomes and Sugden (1982), for example, have shown that the many of these effects follow if one assumes that $r(x,y) = Q(x-y)$ where Q is convex for positive values and concave for negative values.

[47] Although this point in the discussion has been reached by an examination of the preference reversal phenomenon over risky prospects, it is important to note that neither the evidence of response mode effects (e.g., Slovic, 1975) nor their implications for economic analysis are confined to the case of choice under uncertainty.

were true, then with further evidence on exactly *how* the response mode mattered, we could presumably incorporate it into existing theories as a new independent variable. Because response modes tend to be constant within specific economic models (e.g., quantity responses to fixed prices in competitive markets, valuation announcements—truthful or otherwise—in auctions, etc.), we should expect most of the testable implications of this approach to appear as cross-institutional predictions, such as systematic violations of the various equivalency results involving prices versus quantities, or second price/sealed bid versus oral English auctions. I feel that the new results and implications for our theories of institutions and mechanisms would be exciting indeed.[48]

FRAMING EFFECTS

Evidence

In addition to response mode effects, psychologists have uncovered an even more disturbing phenomenon: namely, that alternative means of representing or "framing" probabilistically equivalent choice problems will lead to systematic differences in choice. An early example of this phenomenon is reported by Slovic (1969b), who found, for example, that offering a gain or loss contingent on the joint occurrence of four independent events with probability p elicited responses different from offering it on the occurrence of a single event with probability p^4 (all underlying probabilities were stated explicitly). In comparison with the single-event case, making a gain contingent on the joint occurrence of events was found to make it more attractive; making a loss contingent on the joint occurrence of events made it more unattractive.[49]

In another study, Payne and Braunstein (1971) used pairs of gambles of the type illustrated in Figure 10. Each of the gambles in the figure, known as a *duplex gamble*, involves spinning the pointers on both its "gain wheel" (on the left) and its "loss wheel" (on the right), with the individual

---

[48]A final "twist" on the preference reversal phenomenon: Karni and Safra (1987) and Holt (1986) have shown how the procedures used in most of these studies, namely, the Becker, De-Groot, and Marschak elicitation technique (see footnote 36) and the practice of only selecting a few questions to actually play, will lead to truthful revelation of preferences only under the additional assumption that the individual satisfies the independence axiom. Accordingly, it is possible to construct (and these researchers have done so) examples of non-expected utility individuals with *transitive* underlying preferences and no response mode effects, whose optimal responses in such experiments consist of precisely the typical "preference reversal" responses. How (and whether) experimenters will be able to address this issue remains to be seen.

[49]Even though all underlying probabilities were stated explicitly, Slovic found that individuals tended to overestimate the probabilities of these compound events.

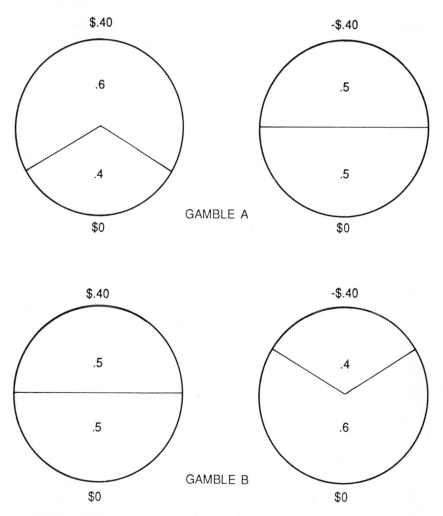

FIGURE 10 Duplex gambles with identical underlying distributions.

face of wealth variations has also been observed in several experimental studies.[51]

Markowitz (1952:155) also suggested that certain circumstances may cause the individual's reference point to deviate temporarily from current wealth. If these circumstances include the manner in which a given problem is verbally described, then differing risk attitudes over gains and losses can

---

[51] See the discussion and references in Machina (1982:285–286).

receiving the sum of the resulting amounts. Thus, an individual choosing gamble A would win $.40 with probability .3 (i.e., if the pointer in the gain wheel landed up and the pointer in the loss wheel landed down), would lose $.40 with probability .2 (if the pointers landed in the reverse positions), and would break even with probability .5 (if the pointers landed either both up or both down). An examination of gamble B reveals that it has an identical underlying distribution; thus, subjects should be indifferent regarding a choice between the two gambles, regardless of their risk preferences. Payne and Braunstein, however, found that individuals in fact chose between such pairs (and indicated nontrivial strengths of preference) in manners that were systematically affected by the attributes of the component wheels. When the probability of winning in the gain wheel was greater than the probability of losing in the loss wheel for each gamble (as in the figure), subjects tended to choose the gamble whose gain wheel yielded the greater probability of a gain (gamble A). In cases in which the probabilities of losing in the loss wheels were greater than the probabilities of winning in the gain wheels, subjects tended to choose the gamble with the lower probability of losing in the loss wheel.

Finally, although the gambles in Figure 10 possess identical underlying distributions, continuity suggests that worsening of the terms of the preferred gamble could result in a pair of nonequivalent duplex gambles in which the individual will actually choose the one with the *stochastically dominated* underlying distribution. In an experiment in which subjects were allowed to construct their own duplex gambles by choosing one from a pair of prospects involving gains and one from a pair of prospects involving losses, stochastically dominated combinations were indeed, chosen (Tversky and Kahneman, 1981; Kahneman and Tversky, 1984).[50]

A second class of framing effects exploits the phenomenon of a *reference point*. Theoretically, the variable that enters an individual's von Neumann-Morgenstern utility functions should be total (i.e., final) wealth, and gambles phrased in terms of gains and losses should be combined with current wealth and reexpressed as distributions over final wealth levels before being evaluated. However, economists since Markowitz (1952) have observed that risk attitudes over gains and losses are more stable than can be explained by a fixed utility function over final wealth, and have suggested that the utility function might be best defined in terms of changes from the "reference point" of current wealth. The stability of risk attitudes in the

---

[50] Subjects were asked to choose either (A) a sure gain of $240 or (B) a 1/4:3/4 chance of $1,000 or $0, and to choose either (C) a sure loss of $750 or (D) a 3/4:1/4 chance of - $1,000 or $0. Eighty-four percent of the subjects chose A over B and 87 percent chose D over C, even though B + C dominates A + D, and choices over the combined distributions were unanimous when they were presented explicitly.

lead to different choices, depending on the exact description. A simple example of this, from Kahneman and Tversky (1979:273), involves the following two questions:

> In addition to whatever you own, you have been given 1,000 (Israeli pounds). You are now asked to choose between a 1/2:1/2 chance of a gain of 1,000 or 0 or a sure chance of a gain of 500.

and

> In addition to whatever you own, you have been given 2,000. You are now asked to choose between a 1/2:1/2 chance of a loss of 1,000 or 0 or a sure loss of 500.

These two problems involve identical distributions over final wealth. When put to two different groups of subjects, however, 84 percent chose the sure gain in the first problem, but 69 percent chose the 1/2:1/2 gamble in the second. A nonmonetary version of this type of example, from Tversky and Kahneman (1981:453), posits the following scenario:

> Imagine that the U.S. is preparing for the outbreak of an unusual Asian disease, which is expected to kill 600 people. Two alternative programs to combat the disease have been proposed. Assume that the exact scientific estimate of the consequences of the programs are as follows:
>
> If Program A is adopted, 200 people will be saved.
>
> If Program B is adopted, there is 1/3 probability that 600 people will be saved, and 2/3 probability that no people will be saved.

Seventy-two percent of the subjects who were presented with this form of the question chose program A. A second group was given the same initial information, but the descriptions of the programs were changed to read (p. 453):

> If Program C is adopted 400 people will die.
>
> If Program D is adopted there is 1/3 probability that nobody will die, and 2/3 probability that 600 people will die.

Although this statement once again implies a problem that is identical to the former one, 78 percent of the respondents chose program D.

In other studies, Schoemaker and Kunreuther (1979); Hershey and Schoemaker (1980b); Kahneman and Tversky (1982, 1984); Hershey, Kunreuther, and Schoemaker (1982); McNeil et al. (1982); and Slovic, Fischhoff, and Lichtenstein (1982) have found that subjects' choices in otherwise identical problems will depend on whether the choices are phrased as decisions about whether to gamble or to insure, whether the statistical information for different therapies is presented in terms of cumulative survival probabilities over time or cumulative mortality probabilities over time, and so forth (see also the additional references in Tversky and Kahneman [1981] as well as the examples of this phenomenon in nonstochastic situations given in Thaler [1980, 1985]).

In a final class of examples, not based on reference point effects, Moskowitz (1974) and Keller (1985) found that the proportion of subjects that choose in conformance with or in violation of the independence axiom in examples like the Allais paradox was significantly affected by whether the problems were described in the standard matrix form (e.g., Raiffa, 1968:7), in a decision tree form, or as minimally structured written statements. Interestingly enough, the form that was judged to be the "clearest representation" by the majority of Moskowitz's subjects (the tree form) led to the lowest degree of consistency with the independence axiom, the highest proportion of Allais-type (i.e., fanning out) choices, and the highest persistency rate of these choices (1974:234, 237–38).

Two Issues Regarding Framing

The replicability and pervasiveness of the above group of examples is indisputable. Their implications for economic modeling involve two issues (at least). The first is whether these experimental observations possess any analogue outside of the laboratory. Real-world decision problems are never as neatly packaged as those that appear on experimental questionnaires; thus, monitoring such effects would not be as straightforward. This difficulty in monitoring does not mean that such efforts do not exist, however, or that they cannot be objectively observed or quantitatively measured. The real-world example that comes most quickly to mind, and is presumably of no small importance to the involved parties, is whether gasoline price differentials should be represented as "cash discounts" or "credit surcharges." Similarly, Russo, Krieser, and Miyashita (1975) and Russo (1977) found that the practice and even the method of displaying unit price information in supermarkets (information that allowed consumers to calculate for themselves) affected both the level and distribution of consumer expenditures. The empirical marketing literature is no doubt replete with findings that could legitimately be interpreted as real-world framing effects.

The second, more difficult issue is that of the independent observability of the particular frame that an individual will adopt in a given problem. In the duplex gamble and matrix/decision tree/written statement examples of the previous section, the different frames seem unambiguously determined by the form of presentation. In instances in which framing involves the choice of a reference point, however, instances that presumably include the majority of real-world cases, this point might not be objectively determined by the form of presentation. Rather, it might be chosen differently and, what is worse, *unobservably*, by each individual.[52] In a particularly thorough and insightful study, Fischhoff (1983) presented subjects with a written decision problem that allowed for different choices of a reference point. The study went on to explore different ways of predicting which frame individuals would adopt in order to be able to predict their actual choices. Although the majority choice of subjects was consistent with what would appear to be the most appropriate frame, Fischhoff noted "the absence of any relation within those studies between [separately elicited] frame preference and option preference." Indeed, to the extent that frame preferences varied across his experiments, they did so inversely to the incidence of the predicted choice (Fischhoff, 1983:115–116).[53] If such problems can occur in predicting responses to specific written questions in the laboratory, imagine how they could plague the modeling of real-world choice behavior.

Framing Effects and Economic Analysis: Has This Problem Already Been Solved?

What response is appropriate if it turns out that framing actually is a real-world phenomenon of economic relevance and, in particular, if individuals' frames cannot always be observed? I would argue that the means of responding to this issue can already be found in the "tool box" of existing economic analysis.

Let us consider first the case in which the frame of a particular economic decision problem (even though it should not matter from the point of view of standard theory), can at least be independently and objectively observed. I believe that, in fact, economists have already solved such a problem in their treatment of the phenomenon of "uninformative advertising." Although it is hard to give a formal definition of this term, it is widely felt that economic theory is hard put to explain a large portion of current advertising in terms of traditional informational considerations.[54] This constraint, however, has hardly led economists to abandon classical consumer theory. Rather, models of uninformative advertising proceed by quantifying this variable (e.g., air time) and treating it as an additional independent variable in the utility function, the demand function, or both. Standard results like the Slutsky equation need not be abandoned but rather reinterpreted as properties of demand functions holding this new variable

---

[52] This is not to say that well-defined reference points never exist. The reference points involved in credit surcharges versus cash discounts, for example, seem unambiguous.

[53] Fischhoff notes that "[i]f one can only infer frames from preferences after assuming the truth of the theory, one runs the risk of making the theory itself untestable" (p. 116).

[54] A wonderful example, offered by my colleague Joel Sobel, is that of milk advertisements that make no reference either to price or to a particular dairy. What commodity could be more well-known than milk?

constant. The degree of advertising itself is determined as a maximizing variable on the part of the firm (given some cost curve) and is thus subject to standard comparative static analysis.

In cases in which decision frames can be observed, framing effects presumably can be modeled in an analogous manner. To do so, one would begin by adopting a method of quantifying—or at least of categorizing—frames. The activity of the second step, some of which has of course already been done, would be to study both the effect of this new independent variable holding the standard economic variables constant, and, conversely, to retest standard economic theories in conditions in which the frame was carefully held in a fixed position. With any luck, one would find that, holding the frame constant, the Slutsky equation still held.

The next step in any given modeling situation would be to discover "who determines the frame." If (as with advertising) it is the firm, then the effect of the frame on consumer demand, and hence on the firm's profits, can be incorporated into the firm's maximization problem. The choice of the frame, as well as the other relevant variables (e.g., prices and quantities), can be simultaneously determined and subjected to comparative static analysis just as in the case of uninformative advertising.

A seemingly more difficult case is when the individual chooses the frame (for example, a reference point), and this choice cannot be observed. Although findings of Fischhoff (1983) should be kept in mind, let us assume that this choice is at least systematic in the sense that the consumer will jointly choose the frame and make the subsequent decision in a way that maximizes a "utility function" that depends both on the decision and on the choice of frame. In other words, individuals make their choices as part of a joint maximization problem, the other component of which (the choice of frame or reference point) cannot be observed.

Such models are hardly new to economic analysis. Indeed, most economic models presuppose that the agent is simultaneously maximizing his or her choices with respect to variables other than the ones being studied. When assumptions are made on the individual's joint preferences over the unobserved and observed variables, the well-known *theory of induced preferences* can be used to derive testable implications on choice behavior over the observables.[55] With a little more knowledge on exactly how frames are chosen, such an approach could presumably be applied here as well.

The above remarks should not be taken as implying that the problems of framing in economic analysis have already been solved or that there is no need to adapt and, if necessary, abandon standard economic models in

---

[55]See, for example, Milne (1981). For an application of the theory of induced preferences to choice under uncertainty, see Machina (1984).

light of this phenomenon. Rather, the remarks reflect the view that when psychologists are able to present enough systematic evidence on how these effects operate, economists will be able to respond appropriately.

OTHER ISSUES: IS PROBABILITY THEORY RELEVANT?

The Manipulation of Subjective Probabilities

The evidence discussed so far has consisted primarily of cases in which subjects were presented with explicit (i.e., "objective") probabilities as part of their decision problems and the models that addressed these phenomena possessed the corresponding property of being defined over objective probability distributions. There is extensive evidence, however, that when individuals have to estimate or revise probabilities for themselves, they will make systematic mistakes in doing so.

The psychological literature on the processing of probabilistic information is much too large even to summarize here. Yet, it is worth noting that experimenters have uncovered several "heuristics" used by subjects that can lead to predictable errors in the formation and manipulation of subjective probabilities. Kahneman and Tversky (1973), Bar-Hillel (1974), and Grether (1980), for example, all found that probability updating systematically departs from Bayes' law in the direction of underweighting prior information and overweighting the "representativeness" of the current sample. In a related phenomenon termed the "law of small numbers," Tversky and Kahneman (1971) found that individuals overestimated the probability of drawing a perfectly representative sample out of a heterogeneous population. Finally, Bar-Hillel (1973), Tversky and Kahneman (1983), and others have found systematic biases in the formation of the probabilities of conjunctions of both independent and nonindependent events. For surveys, discussions, and examples of the psychological literature on the formation and handling of probabilities, see Edwards, Lindman, and Savage (1963); Edwards (1971); Slovic and Lichtenstein (1971); Tversky and Kahneman (1974); and Grether (1978), as well as the collections in *Acta Psychologica* (December 1970); Kahneman, Slovic, and Tversky (1982); and Arkes and Hammond (1986). For examples of how economists have responded to some of these issues, see Arrow (1982), Viscusi (1985a,b) and the references cited there.

The Existence of Subjective Probabilities

The evidence referred to above indicates that when individuals are asked to formulate probabilities they seldom do it correctly. These findings may be rendered moot, however, by evidence that suggests that when individuals making decisions under uncertainty are not explicitly asked to form subjective probabilities they might not do it at all.

In one of a class of examples developed by Ellsberg (1961), subjects were presented with a pair of urns: the first contained 50 red balls and 50 black balls, and the second also contained 100 red and black balls but in an unknown proportion. When faced with the choice of staking a prize on (R_1) drawing a red ball from the first urn, (R_2) drawing a red ball from the second urn, (B_1) drawing a black ball from the first urn, or (B_2) drawing a black ball from the second urn, a majority of subjects strictly preferred (R_1) over (R_2) and strictly preferred (B_1) over (B_2). It is clear that there can exist no subjectively assigned of probabilities $p : (1 - p)$ of drawing a red versus a black ball from the second urn—not even 1/2:1/2, that can simultaneously generate both of these strict preferences. Similar behavior in this and related problems has been observed by Raiffa (1961), Becker and Brownson (1964), MacCrimmon (1965), Slovic and Tversky (1974), and MacCrimmon and Larsson (1979).[56]

Life (and Economic Analysis) Without Probability Theory

One response to this type of phenomenon as been to suppose that individuals "slant" whatever subjective probabilities they might otherwise form in a manner that reflects the amount of confidence or ambiguity associated with them (Fellner, 1961, 1963; Becker and Brownson, 1964; Brewer and Fellner, 1965; Fishburn, 1985, 1986; Hogarth and Kunreuther, 1985, 1986; and Einhorn and Hogarth, 1986). In the case of complete ignorance regarding probabilities, Arrow and Hurwicz (1972), Maskin (1979), and others have presented axioms that imply such principles as ranking options solely on the basis of their best or worst possible outcomes or (both) (e.g., maximin, maximax), the unweighted average of their outcomes ("principle of insufficient reason"), or similar criteria.[57] Finally generalizations of expected utility theory that drop the standard additivity or compounding laws of probability theory (or both) have been developed by Schmeidler (1989) and Segal (1987).

Although the above models may well capture aspects of actual decision processes, analytically the most useful approach to choice in the presence of uncertainty but the absence of probabilities is the so-called *state-preference model* of Arrow (1953/1964), Debreu (1959), and Hirshleifer (1965, 1966).[58]

---

[56] See also the discussions of Fellner (1961, 1963), Brewer (1963), Ellsberg (1963), Roberts (1963), Brewer and Fellner (1965), MacCrimmon (1968), Smith (1969), Sherman (1974), and Sinn (1980).

[57] For an excellent discussion of the history, nature, and limitations of such approaches, see Arrow (1951).

[58] For a comprehensive overview of this model and its analytics, see Karni (1985).

In this model, uncertainty is represented by a set of mutually exclusive and exhaustive *states of nature* $S = \{s_i\}$. This partition of all possible unfoldings of the future could be either coarse, such as the pair of states {it rains here tomorrow, it does not rain here tomorrow}, or else very fine (so that the definition of a state might read "it rains here tomorrow *and* the temperature at Gibraltar is 75 degrees at noon *and* the price of gold in London is below \$700 per ounce"). Note that it is neither feasible nor desirable to capture all conceivable sources of uncertainty when specifying the set of states for a given problem. It is not feasible because no matter how finely the states are defined, there will always be some other random criterion on which to further divide them; it is not desirable because such criteria may affect neither individuals' preferences nor their opportunities. Rather, the key requirements are that the states be mutually exclusive and exhaustive so that exactly one will be realized, and that the extent to which the individual is able to influence their probabilities (if at all) be explicitly specified.

Given a fixed (and, let us say, finite) set of states, the objects of choice in this framework consist of alternative *state-payoff bundles*, each of which specifies the outcome the individual will receive in every possible state. When, for example, the outcomes are monetary payoffs, state-payoff bundles take the form $(c_1, \ldots, c_n)$, where c_i denotes the payoff the individual will receive should state i occur. In the case of exactly two states of nature, this set can be represented set by the points in the (c_1, c_2) plane. Because bundles of the form (c, c) represent prospects that yield the same payoff in each state of nature, the 45-degree line in this plane is known as the *certainty line*.

Now, if the individual happens to assign some set of probabilities $\{p_i\}$ to the states $\{s_i\}$, each bundle $(c_1, \ldots, c_n)$ will imply a specific probability distribution over the payoffs, and his or her preferences could be inferred (i.e., indifference curves) over state-payoff bundles.[59] Yet because these bundles are defined directly over the respective states and without reference to any probabilities, it is possible to speak of preferences over such bundles without making any assumptions regarding the coherency, or even the existence, of probabilistic beliefs. Researchers such as those listed above, as well as Yaari (1969), Diamond and Yaari (1972), and Mishan (1976), have used this indifference curve-based approach to derive results from individual demand behavior through general equilibrium in a context that requires

---

[59] In generating these indifference curves from individuals' preferences over probability distributions, one implicitly assumes that their level of satisfaction from a given amount of money does not depend on the particular state of nature which that (i.e., that their preferences are state independent. Beginning with the next sentence, this assumption will no longer be required.

neither the expected utility hypothesis nor the existence or commonality of subjective probabilities. In other words, life without probability theory does not imply life without economic analysis.[60]

IMPLICATIONS FOR PRIVATE AND PUBLIC DECISION MAKING

Fifteen years ago, a decision analyst who was advising an individual, firm, or government agency in a choice under uncertainty might use something like the following stylized procedure.

1. Collect as much information as possible about the decision, and construct an explicit list of the currently (and potentially) available options.

2. Assess the decision maker's (or, alternatively, the "experts'") subjective probability distributions over consequences implied by each option.

3. Evaluate the decision maker's (or, alternatively, "society's') preferences regarding the alternative consequences, including their attitudes toward risk (in other words, assess their von Neumann-Morgenstern utility function).

4. Determine the option that would yield the highest (individual or social) expected utility.[61]

Of course, the consequences might involve several dimensions (requiring the assessment of a multivariate utility function),[62] or the experts might disagree on the probabilities (requiring some form of consensus, aggregation, or pooling of beliefs).[63] Nevertheless, researchers working on these aspects remained confident of the validity of this overall (expected utility-based) approach.

Should the developments surveyed in this paper change the way private decision analysts or public decision makers go about their jobs? Do they imply new or different business or governmental responsibilities in keeping customers or citizens informed of any voluntary (or involuntary) risks they

---

[60] A final issue is the lack of a unified model that is capable of simultaneously handling all of the phenomena described in this paper: fanning out, the preference reversal phenomenon, framing effects, probability biases, and the Ellsberg paradox. After all, it is presumably the same individuals who are *exhibiting* each of these phenomena; should there not be a single model capable of *generating* them all? I doubt whether our current ability allows us to do this; I also doubt the need for a unified model as a prerequisite for further progress. The aspects of behavior considered in this paper are quite diverse, and if (like the wave-versus-particle properties of light) they cannot be currently unified, this does not mean that we cannot continue to learn by studying and modeling them separately.

[61] The classic introductory expositions of the process of decision analysis are Raiffa (1968) and Schlaifer (1969).

[62] See, for example, Keeney and Raiffa (1976).

[63] See, for example, Grofman and Owen (1986).

may be facing? The following section discusses some of the issues that these new developments raise.

Implications for Private-Sector Decision Analysis

How should private-sector decision analysts adapt their procedures in light of these new empirical findings and theoretical models? It is hard to see how step 1 (formulating the options) could or should change. Yet the types of systematic biases in the formulation and manipulation of subjective probabilities presented in the preceding section ("Other Issues: Is Probability Theory Relevant?") should cause the analyst to be especially careful in obtaining mutually consistent estimates of the underlying event likelihoods used in constructing the probability distributions over consequences implied by each option in step 2. Note that this step has nothing to do with the client's attitudes toward bearing these risks (i.e., whether or not they do, or should, maximize expected utility). Rather, it consists of applying probability theory to establish the internal consistency and (once that has been established) the logical implications of the client's or experts' probabilistic beliefs. If the client assigns probability .3 to the occurrence of some event A, probability .2 to the occurrence of some mutually exclusive event B, and probability .6 to the occurrence of neither, then at least one of these numbers will have to change before the pieces will fit. This situation is no different from that of asking a client for the length, width, and area of his or her living room before offering advice on a choice of carpet: if the numbers do not multiply out correctly then something is wrong, and the advising process should stop short until corrections are made. Although I suspect practitioners in the field have been aware of such inconsistencies (and of how to "iron them out") for some time now, the type of systematic and specific biases that psychologists have been uncovering now give decision analysts the opportunity, and, I feel, much more of an obligation, to search explicitly for and eliminate biases and inconsistencies in clients' probabilistic beliefs which might otherwise remain hidden.

Although I feel the suggestions of the previous paragraph are important, they are more of a technical improvement than a basic change in how step 2 is carried out. On the other hand, I would argue that the developments reviewed in this paper *do* imply a fundamental change in the way modern decision analysts should proceed with steps 3 and 4 (explicating clients' risk preferences and determining their optimal action). The classical approach would be essentially to impose the property of linearity in the probabilities on the client by assessing his or her von Neumann-Morgenstern utility function and then using it to calculate the "optimal" (i.e., expected utility-maximizing) choice. If clients made choices like those in the Allais paradox, the common consequence effect, or the common ratio effect discussed earlier, or their responses to alternative assessment methods yielded different "recovered" utility functions, they would often be told that they had "inconsistent" (i.e., not expected utility) preferences that would have to be corrected before their optimal action could be determined.

Although experimental subjects and real-world decision makers sometimes do make mistakes in expressing their preferences, I feel that the widespread and systematic nature of "fanning out"-type departures from expected utility, and the growing number of models that can simultaneously accommodate this phenomenon, as well as the more traditional properties of stochastic dominance preference and risk aversion, increase both the analyst's ability and obligation to fit and represent clients' risk attitudes within a consistent non-expected utility framework when their expressed risk preferences are pointing in that direction.[64] Why do I feel that departures from the strictures of probability theory should be corrected but that (systematic) departures from the strictures of expected utility theory should not? Because the former involve the determination of the risks involved in an option, which is a matter of accurate representation, whereas the latter involve the client's willingness to bear these risks, which is a matter of preference. To continue my earlier analogy, reporting a length, width, and area of a room that are not commensurate implies an internally inconsistent description of the room and is simply wrong; preferring purple polka-dot carpeting, however, is a matter of clients' tastes, to which they have every right if it is their living room. In the case of health or environmental risks, this would correspond to the distinction between measuring the detrimental effects of a drug or a pollutant versus determining the individual patient's or society's attitudes toward bearing these consequences.

Does this increased respect for clients' preferences mean that the decision analyst should not play *any* guiding role in steps 3 or 4? The answer is no: conscientious decision analysts will still try to elicit and explicitly represent the client's risk attitudes, their underlying properties (e.g., whether they are risk averse, linear in the probabilities, etc.), and their logical implications. Even more important, they will continue explicitly to separate their client's *beliefs* from their *preferences*. For example, let us say that, although option 1 offers a very high chance of an acceptable but not terrific outcome (e.g., amputation of a gangrenous limb), the client insists on "optimism" or "wishful thinking" in connection with option 2 (e.g., drug therapy), which is not as likely to succeed but does offer a small chance of obtaining the best possible outcome. In that case the decision analysis should take pains formally to represent the client's attitude as a

---

[64]The components of such models (e.g., the functions $\nu(\cdot)$, $\pi(\cdot)$, $\tau(\cdot)$, and $g(\cdot)$ in Table 1) can be assessed by procedures similar to the one described earlier for von Neumann-Morgenstern utility functions.

willingness to bear risk (either by a convex utility function, such as as in Figure 1b, or by some non-expected utility counterpart) rather than as an exaggerated probability estimate of obtaining the best outcome under option 2. The job of the decision analyst has hardly become obsolete.

Implications for Public Decision Making

Although private-sector decision analysts typically act on behalf of an individual client or firm, the decision maker in federal, state, or local government is faced with the obligation of acting on the behalf of citizens whose preferences and interests will generally differ from one another. In the case of decisions under certainty, economists have developed a large body of techniques, collectively termed *welfare economics* or *welfare analysis*, with which to analyze such situations.[65] Not surprisingly, economic theorists have also used the expected utility model as a framework for extending such analyses to a world of uncertainty (e.g., Arrow, 1953/1964, 1974; Diamond, 1967). Let us say, however, that we wish to respect what the recent evidence implies about individuals' actual attitudes toward risk. Can classical welfare analysis, the economist's most important tool for formal policy evaluation, be undertaken with these newer models of preferences?

The answer to this question depends on the model. Fanning-out behavior and the non-expected utility models used to characterize it, as well as the state-payoff approach discussed earlier, are completely consistent with the assumption of well-defined, transitive individual preference orderings and hence with traditional welfare analysis along the lines of Pareto (1909), Bergson (1938) and Samuelson (1947/1983:Chap. 8). For example, the proof of the general efficiency ("Pareto efficiency") of a system of complete contingent-commodity markets (Arrow, 1953/1964; Debreu, 1959:Chap. 7) requires neither the expected utility hypothesis nor the assumption of well-defined probabilistic beliefs. On the other hand, it is clear that the preference reversal phenomenon and framing effects, and at least some of the nontransitive or noneconomic models used to address them, will prove much more difficult to reconcile with welfare analysis, or at least with welfare analysis as currently practiced.

To see how some of these (transitive) non-expected utility models can be applied to policy questions, the reader may recall the earlier expected utility-based analysis of the trade-off between the probability p and magnitude L of a disastrous event (see the section entitled "The Expected Utility Model"). Under the expected utility hypothesis, it was seen that an individual's marginal rate of substitution (i.e., acceptable rate of trade-off)

---

[65] The standard policy techniques of "benefit-cost analysis," "benefit-risk analysis," and so forth fall into this category.

between these variables would vary exactly proportionally to the loss probability p (as seen in footnote 11). Although it requires a bit of algebra to do so, it is possible to demonstrate that if preferences depart from expected utility by "fanning out" (Figures 4b, 5a, 5b, and 6), then individuals' marginal rates of substitution between p and L will always vary *less than proportionally* to the loss probability p (Machina, 1983b:282–289). Although this is not as strong a prediction as expected utility theory's prediction of exact proportionality, it can be used to place at least a one-sided bound on how individuals' acceptable rates of trade-off behave. In any event, it is at least more closely tied to what has actually been observed about preferences over risky prospects.[66]

Public and Corporate Obligations in the Presentation of Information

The final issue concerns the public policy implications of framing effects. If individuals' choices actually depend on the manner in which publicly or privately supplied probabilistic information (e.g., cancer incidences or flood probabilities) is presented, then the manner of presentation *itself* becomes a public policy issue over which interest groups may well contend. Should "freedom of information" imply that a government or manufacturer has an obligation to present a broad range of "legitimate" frames when disclosing required information, or would this practice lead to confusion and waste? Should legal rights of recourse for failures to provide information (e.g., job or product hazards) extend to failures to frame it "properly"? The general issue of public perception of risk is of growing concern to a number of government agencies—in particular, the Environmental Protection Agency.[67] To the extent that new products, medical techniques, and environmental hazards continue to appear and the government takes a role their regulation, these issues will become more and more pressing.

Although the issue of the public and private framing of probabilistic information is a comparatively new one, I feel that there are several

---

[66] In the nuclear power plant example the earlier section noted above, let us say that there was a probability p of an accident involving a loss of L, that some expected utility maximizer was just willing to accept an increase of ΔL_1 in this potential loss in return for a reduction of Δp in its probability, and that some individual with fanning-out preferences was just willing to accept an increase of ΔL_2 in return for this same reduction in p (ΔL_2 could be greater or less than ΔL_1). Should new technology reduce the initial loss probability by half to $p/2$, the extra loss the expected utility maximizer would be willing to accept for a (further) reduction of Δp would also drop by half (i.e., to $\Delta L_1/2$. The extra loss that the individual with fanning-out preferences would be willing to accept, however, would drop by *less* than half (i.e., to some amount *greater* than $\Delta L_2/2$. In other words, as the probability of the accident drops, the individual with fanning-out preferences will exhibit a comparatively greater willingness to trade off increases in loss magnitude in exchange for further reductions in loss probability.

[67] See Russell (in this volume).

analogous issues (not all of them fully resolved) from which useful insights may be derived. Previous examples have included the cash discount/credit surcharge issue mentioned earlier, rotating warning labels on cigarette packages, financial disclosure regulations, bans on certain forms of alcohol advertising, publicity requirements for product recall announcements, and current debate cover such issues as requiring special labels on irradiated produce or on products imported from countries that engage in human or animal rights violations, or both. If these issues do not provide ready-made answers for the case of probabilistic information, they at least allow a glimpse of how policy makers, interest groups, and the public feel and act toward the general issue of the presentation of information.

ACKNOWLEDGMENT

This paper is an adapted and expanded version of Machina (1987). I am grateful to Brian Binger, John Conlisk, Rob Coppock, Jim Cox, Vince Crawford, Gong Jin-Dong, Elizabeth Hoffman, Brett Hammond, Paul Portney, Howard Raiffa, Michael Rothschild, Carl Shapiro, Vernon Smith, Joseph Stiglitz, Timothy Taylor, Richard Thaler, and especially Joel Sobel for helpful discussions on this material.

EDITORS' NOTE

In the preceding paper, Mark Machina addresses an important set of questions relevant to the use of benefit-cost analysis in regulatory decision making. These include:

• How do (and should) decision makers and analysts cope with uncertainties in the science underlying decisions?

• How does the way in which information is presented—or framed— affect analysis and regulatory decisions?

• Is it possible to get better estimates of uncertainty for purposes of practical decision making?

Machina examines recent theoretical and empirical findings on how people actually evaluate risks and assign probabilities in arriving at policy preferences. Much of this literature challenges the traditional economic approach to preferences, in that it points out that individuals are able to sustain nonlinearity in their subjective assessments of probabilities, reversal of preferences over time or between different situations, and differences in preferences and probability assessments depending on the way in which a problem (or analysis) is framed or presented.

At the heart of his paper is the conclusion that predicted policy outcomes should be differentiated from individual and collective policy preferences for purposes of analysis and decision making. As he puts it, observed "departures from the strictures of probability theory should be corrected [by the analyst or the decision maker] but that [systematic] departures from the structures of expected utility theory should not." This is because the former involved the determination of the risks associated with alternative actions or policies, which are in fact matters of accurate representation, while the latter involve the willingness of individuals, organizations, and society to bear these risks, which is a matter of preference.

He concludes that analysis must be designed to account for actual preferences, even those that depart from the tenets of expected utility theory. Therefore, analysts and decision makers, in assigning values to policy alternatives, may need to consider departures from expected utility and weighting schemes to reflect those departures.

REFERENCES

Akerlof, G.
   1970   The market for "Lemons": Quality uncertainty and the market mechanism. *Quarterly Journal of Economics* 84:488–500.
   1984   *An Economic Theorist's Book of Tales.* Cambridge: Cambridge University Press.
Allais, M.
   1952   Fondements d'une théorie positive des choix comportant un risque et critique des postulats et axiomes de l'ecole Américaine. *Econométrie*, Colloques Internationaux du Centre National de la Recherche Scientifique 40:Paris, 1953, 257–332.
   1953   Le comportement de l'homme rationel devant le risque, Critique des postulates et axiomes de l'ecole Américaine. *Econometrica* 21:503–546. Summarized version of Allais (1952).
   1979a   The foundations of a positive theory of choice involving risk and a criticism of the postulates and axioms of the American school. Pp. 27–145 in M. Allais and O. Hagen, eds. *Expected Utility Hypotheses and the Allais Pardox.* Dordrecht, Holland: D. Reidel Publishing Company.
   1979b   The so-called Allais paradox and rational decisions under uncertainty. Pp. 437–681 in M. Allais and O. Hagen, eds. *Expected Utility Hypotheses and the Allais Pardox.* Dordrecht, Holland: D. Reidel Publishing Company.
Allen, B.
   1987   Smooth preferences and the local expected utility hypothesis. *Journal of Economic Theory* 41:340–355.
Arrow, K.
   1951   Alternative approaches to the theory of choice in risk-taking situations. *Econometrica* 19:404–437.
   1953   Le role des valeurs boursières pour la répartition le meilleure des risques. *Econométrie*, Colloques Internationaux du Centre National de la Recherche Scientifique 40, Paris, 1953, 41–47.
   1963   Comment. *Review of Economics and Statistics* 45(Suppl.):24–27.
   1964   The role of securities in the optimal allocation of risk-bearing. *Review of Economic Studies* 31:91–96.
   1974   *Essays in the Theory of Risk-Bearing.* Amsterdam: North-Holland Publishing Company.

1982   Risk perception in psychology and economics. *Economic Inquiry* 20:1–9.
Arrow, K., and L. Hurwicz
1972   An optimality criterion for decision-making under ignorance. Pp. 1–11 in Carter,
           D., and F. Ford, eds., *Uncertainty and Expectations in Economics*. Oxford: Basil
           Blackwell.
Arrow, K., and M. Intriligator, eds.
1981   *Handbook of Mathematical Economics, vol. 1.* Amsterdam: North-Holland
           Publishing Company.
Bar-Hillel, M.
1973   On the subjective probability of compound events. *Organizational Behavior and
           Human Performance* 9:396–406.
1974   Similarity and probability. *Organizational Behavior and Human Performance*
           11:277–282.
Battalio, R., J. Kagel, and D. MacDonald
1985   Animals' choices over uncertain outcomes. *American Economic Review* 75:597–
           613.
Becker, G., M. DeGroot, and J. Marschak
1964   Measuring utility by a single-response sequential method. *Behavioral Science*
           9:226–232.
Becker, S., and F. Brownson
1964   What price ambiguity? Or the role of ambiguity in decision-making. *Journal of
           Political Economy* 72:62–73.
Bell, D.
1982   Regret in decision making under uncertainty. *Operations Research* 30:961–980.
1983   Risk premiums for decision regret. *Management Science* 29:1066–1076.
1985   Disappointment in decision making under uncertainty. *Operations Research*
           33:1–27.
Bell, D., and H. Raiffa
1980   Decision Regret: A Component of Risk Aversion. Unpublished manuscript,
           Harvard University.
Berg, J., J. Dickhaut, and J. O'Brien
1983   Preference Reversal and Arbitrage. Unpublished manuscript, University of
           Minnesota.
Bergson, A.
1938   A reformulation of certain aspects of welfare economics. *Quarterly Journal of
           Economics* 52:310–334.
Bernoulli, D.
1738   Specimen theoriae novae de mensura sortis. *Commentarii Academiae Scien-
           tiarum Imperialis Petropolitanae* [Papers of the Imperial Academy of Sciences in
           Petersburg] 5:175–192. English translation: Exposition of a new theory on the
           measurement of risk. *Econometrica* 22(1954):23–36.
Blyth, C.
1972   Some probability paradoxes in choice from among random alternatives. *Journal
           of the American Statistical Association* 67:366–373.
Brewer, K.
1963   Decisions under uncertainty: Comment. *Quarterly Journal of Economics* 77:159–
           161.
Brewer, K., and W. Fellner
1965   The slanting of subjective probabilities—Agreement on some essentials. *Quarterly
           Journal of Economics* 79:657–663.

Chew, S.
  1983   A generalization of the quasilinear mean with applications to the measurement of income inequality and decision theory resolving the Allais Paradox. *Econometrica* 51:1065–1092.
Chew, S., and K. MacCrimmon
  1979a   Alpha-Nu Choice Theory: A Generalization of Expected Utility Theory. University of British Columbia Faculty of Commerce and Business Administration Working Paper No. 669.
  1979b   Alpha Utility Theory, Lottery Composition, and the Allais Paradox. University of British Columbia Faculty of Commerce and Business Administration Working Paper No. 686.
Chew, S., and W. Waller
  1986   Empirical tests of weighted utility theory. *Journal of Mathematical Psychology* 30:55–72.
Chew, S., E. Karni, and Z. Safra
  1987   Risk aversion in the theory of expected utility with rank dependent probabilities. *Journal of Economic Theory* 42:370–381.
Davis, J.
  1958   The transitivity of preferences. *Behavioral Science* 3:26–33.
Debreu, G.
  1959   *Theory of Value: An Axiomatic Analysis of General Equilibrium.* New Haven: Yale University Press.
Dekel, E.
  1986   An axiomatic characterization of preferences under uncertainty: Weakening the independence axiom. *Journal of Economic Theory* 40:304–318.
Diamond, P.
  1967   The role of a stock market in a general equilibrium model with technological uncertainty. *American Economic Review* 57:759–773.
Diamond, P., and M. Rothschild, eds.
  1978   *Uncertainty in Economics.* New York: Academic Press.
Diamond, P., and M. Yaari
  1972   Implications of the theory of rationing for consumer choice under uncertainty. *American Economic Review* 62:333–343.
Edwards, W.
  1954a   Probability-preferences among bets with differing expected value. *American Journal of Psychology* 67:56–67.
  1954b   The theory of decision making. *Psychological Bulletin* 51:380–417.
  1955   The prediction of decisions among bets. *Journal of Experimental Psychology* 50:201–214.
  1962   Subjective probabilities inferred from decisions. *Psychological Review* 69:109–135.
  1971   Bayesian and regression models of human information processing—A myopic perspective. *Organizational Behavior and Human Performance* 6:639–648.
Edwards, W., H. Lindman, and L. Savage
  1963   Bayesian statistical inference for psychological research. *Psychological Review* 70:193–242.
Einhorn, H., and R. Hogarth
  1986   Decision making under ambiguity. *Journal of Business* 59(Suppl.):S225–S250.
Ellsberg, D.
  1961   Risk, ambiguity, and the Savage axioms. *Quarterly Journal of Economics* 75:643–669.

1963 Risk, ambiguity, and the Savage axioms: Reply. *Quarterly Journal of Economics* 77:336–342.

Epstein, L.
1985 Decreasing risk aversion and mean-variance analysis. *Econometrica* 53:945–961.
1987 The unimportance of the intransitivity of separable preferences. *International Economic Review* 28:315–322.

Feather, N.
1959 Subjective probability and decision under uncertainty. *Psychological Review* 66:150–164.

Fellner, W.
1961 Distortion of subjective probabilities as a reaction to uncertainty. *Quarterly Journal of Economics* 75:670–689.
1963 Slanted subjective probabilities and randomization: Reply to Howard Raiffa and K. R. W. Brewer. *Quarterly Journal of Economics* 77:676–690.

Fischhoff, B.
1983 Predicting frames. *Journal of Experimental Psychology: Learning, Memory and Cognition* 9:103–116.

Fishburn, P.
1981 An axiomatic characterization of skew-symmetric bilinear functionals, with applications to utility theory. *Economics Letters* 8:311–313.
1982 Nontransitive measurable utility. *Journal of Mathematical Psychology* 26:31–67.
1982 Transitive measurable utility. *Journal of Economic Theory* 31:293–317.
1984a SSB utility theory: An economic perspective. *Mathematical Social Sciences* 8:63–94.
1984b SSB utility theory and decision making under uncertainty. *Mathematical Social Sciences* 8:253–285.
1986 A new model for decisions under uncertainty. *Economics Letters* 21:127–130.
1988 Uncertainty aversion and separated effects in decision making under uncertainty. In J. Kacprzyk and M. Fedrizzi, eds., *Combining Fuzzy Impressions With Probabilistic Uncertainty in Decision Making.* Berlin: Springer Verlag.

Fishburn, P., and G. Kochenberger
1979 Two-piece Von Neumann-Morgenstern utility functions. *Decision Sciences* 10:503–518.

Friedman, M., and L. Savage
1948 The utility analysis of choices involving risk. *Journal of Political Economy* 56:279–304. Reprinted with revisions in Stigler and Boulding (1952).

Grether, D.
1978 Recent psychological studies of behavior under uncertainty. *American Economic Review Papers and Proceedings* 68:70–74.
1980 Bayes rule as a descriptive model: The representativeness heuristic. *Quarterly Journal of Economics* 95:537–557.

Grether, D., and C. Plott
1979 Economic theory of choice and the preference reversal phenomenon. *American Economic Review* 69:623–638.

Grofman, B., and G. Owen, eds.
1986 *Information Pooling and Group Decision Making.* Greenwich, Conn.: JAI Press.

Hagen, O.
1979 Towards a positive theory of preferences under risk. Pp. 271–302 in M. Allais and O. Hagen, eds. *Expected Utility Hypotheses and the Allais Paradox.* Dordrecht, Holland: D. Reidel Publishing Company.

Henderson, J., and R. Quandt
   1980   *Microeconomic Theory: A Mathematical Approach*, 3d ed. New York: McGraw-Hill.
Hershey, J., and P. Schoemaker
   1980a   Prospect theory's reflection hypothesis: A critical examination. *Organizational Behavior and Human Performance* 25:395–418.
   1980b   Risk-taking and problem context in the domain of losses—An expected utility analysis. *Journal of Risk and Insurance* 47:111–132.
   1985   Probability versus certainty equivalence methods in utility measurement: Are they equivalent? *Management Science* 31:1213–1231.
Hershey, J., H. Kunreuther, and P. Schoemaker
   1982   Sources of bias in assessment procedures for utility functions. *Management Science* 28:936–954.
Hey, J.
   1979   *Uncertainty in Microeconomics.* Oxford: Martin Robinson and Company, Ltd.
Hirshleifer, J.
   1965   Investment decision under uncertainty: Choice-theoretic approaches. *Quarterly Journal of Economics* 79:509–536.
   1966   Investment decision under uncertainty: Applications of the state-preference approach. *Quarterly Journal of Economics* 80:252–277.
Hogarth, R.
   1975   Cognitive processes and the assessment of subjective probability distributions. *Journal of the American Statistical Association* 70:271–289.
   1980   *Judgment and Choice: The Psychology of Decision.* New York: John Wiley and Sons.
Hogarth, R., and H. Kunreuther
   1986   Risk, Ambiguity and Insurance. Unpublished manuscript, University of Chicago.
   1989   Risk, ambiguity and insurance. *Journal of Risk and Uncertainty* 2:5–35.
Hogarth, R., and M. Reder, eds.
   1987   *Rational Choice: The Contrast Between Economics and Psychology.* Chicago: University of Chicago Press.
Holt, C.
   1986   Preference reversals and the independence axiom. *American Economic Review* 76:508–515.
Kahneman, D., and A. Tversky
   1973   On the psychology of prediction., *Psychological Review* 80:237–251.
   1979   Prospect theory: An analysis of decision under risk. *Econometrica* 47:263–291.
   1982   The psychology of preferences. *Scientific American* 246:160–173.
   1984   Choices, values and frames. *American Psychologist* 39:341–350.
Kahneman, D., P. Slovic, and A. Tversky, eds.
   1982   *Judgment Under Uncertainty: Heuristics and Biases.* Cambridge: Cambridge University Press.
Karmarkar, U.
   1974   The Effect of Probabilities on the Subjective Evaluation of Lotteries. Massachusetts Institute of Technology Sloan School of Business Working Paper No. 698-74.
   1978   Subjectively weighted utility: A descriptive extension of the expected utility model. *Organizational Behavior and Human Performance* 21:61–72.
   1979   Subjectively weighted utility and the Allais paradox. *Organizational Behavior and Human Performance* 24:67–72.

Karni, E.
1985   *Decision Making Under Uncertainty: The Case of State Dependent Preferences.*
        Cambridge, Mass.: Harvard University Press.
Karni, E., and Z. Safra
1987   "Preference reversal" and the observability of preferences by experimental
        methods. *Econometrica* 55:675–685.
Keeney, R., and H. Raiffa
1976   *Decisions with Multiple Objectives: Preferences and Value Tradeoffs.* New York:
        John Wiley and Sons.
Keller, L.
1985   The effects of decision problem representation on utility conformance. *Manage-
        ment Science* 6:738–751.
Kihlstrom, R., D. Romer, and S. Williams
1981   Risk aversion with random initial wealth. *Econometrica* 49:911–920.
Kim, T., and M. Richter
1986   Nontransitive-nontotal consumer theory. *Journal of Economic Theory* 38:324–363.
Knetsch, J., and J. Sinden
1984   Willingness to pay and compensation demanded: Experimental evidence of
        an unexpected disparity in measures of value. *Quarterly Journal of Economics*
        99:507–521.
Knez, M., and V. Smith
1987   Hypothetical valuations and preference reversals in the context of asset trading.
        Pp. 131–154 in A. Roth, ed., *Laboratory Experiments in Economics: Six Points
        of View.* Cambridge: Cambridge University Press.
Lichtenstein, S., and P. Slovic
1971   Reversals of preferences between bids and choices in gambling decisions. *Journal
        of Experimental Psychology* 89:46–55.
1973   Response-induced reversals of preference in gambling: An extended replication
        in Las Vegas. *Journal of Experimental Psychology* 101:16–20.
Lindman, H.
1971   Inconsistent preferences among gambles. *Journal of Experimental Psychology*
        89:390–397.
Lippman, S., and J. McCall
1981   The economics of uncertainty: Selected topics and probabilistic methods. Pp.
        211–284 in K Arrow and M. Intriligator, eds., *Handbook of Mathematical
        Economics*, vol. 1. Amsterdam: North-Holland Publishing Company.
Loomes, G., and R. Sugden
1982   Regret theory: An alternative theory of rational choice under uncertainty.
        Economic Journal 92:805–824.
1983a   Regret theory and measurable utility. *Economics Letters* 12:19–22.
1983b   A rationale for preference reversal. *American Economic Review* 73:428–432.
MacCrimmon, K.
1965   An Experimental Study of the Decision Making Behavior of Business Executives.
        Doctoral dissertation, University of California, Los Angeles.
1968   Descriptive and normative implications of the decision-theory postulates. Pp.
        3–23 in K Borch and J. Mossin, *Risk and Uncertainty: Proceedings of a
        Conference Held by the International Economic Association.* London: Macmillan
        and Company.

MacCrimmon, K., and S. Larsson
1979   Utility theory: Axioms versus "Paradoxes." Pp. 333–409 in M. Allais and O. Hagen, eds., *Expected Utility Hypotheses and the Allais Paradox.* Dordrecht, Holland: D. Reidel Publishing Company.

MacCrimmon, K., and D. Wehrung
1986   *Taking Risks: The Management of Uncertainty.* New York: The Free Press.

Machina, M.
1982   "Expected utility" analysis without the independence axiom. *Econometrica* 50:277–323.

1983a   The Economic Theory of Individual Behavior Toward Risk: Theory, Evidence and New Directions. Stanford University Institute for Mathematical Studies in the Social Sciences Technical Report No. 443.

1983b   Generalized expected utility analysis and the nature of observed violations of the independence axiom. Pp. 263–293 in B. Stigum and F. Wenstop, *Foundations of Utility and Risk Theory with Applications.* Dordrecht, Holland: D. Reidel Publishing Company.

1984   Temporal risk and the nature of induced preferences. *Journal of Economic Theory* 33:199–231.

1987   Choice under uncertainty: Problems solved and unsolved. *Journal of Economic Perspectives* 1:121–154.

Machina, M., and W. Neilson
1987   The Ross measure of risk aversion: Strengthening and extension. *Econometrica* 55:1139–1149.

Markowitz, H.
1952   The utility of wealth. *Journal of Political Economy* 60:151–158.

1959   *Portfolio Selection: Efficient Diversification of Investments.* New Haven, Conn.: Yale University Press.

Marschak, J.
1950   Rational behavior, uncertain prospects, and measurable utility. *Econometrica* 18:111–141 (Errata. *Econometrica* 18:312).

1974   *Economic Information, Decision, and Prediction,* 3 vols. Dordrecht, Holland: D. Reidel Publishing Company.

Mas-Colell, A.
1974   An equilibrium existence theorem without complete or transitive preferences. *Journal of Mathematical Economics* 3:237–246.

Maskin, E.
1979   Decision making under ignorance with implications for social choice. *Theory and Decision* 11:319–337.

May, K.
1954   Intransitivity, utility, and the aggregation of preference patterns. *Econometrica* 22:1–13.

McCord, M., and R. de Neufville
1983   Empirical demonstration that expected utility analysis is not operational. Pp. 181–199 in B. Stigum and F. Wenstøp, *Foundations of Utility and Risk Theory with Applications.* Dordrecht, Holland: D. Reidel Publishing Company.

McCord, M., and R. de Neufville
1984   Utility dependence on probability: An empirical demonstration. *Large Scale Systems* 6:91–103.

McNeil, B., S. Pauker, H. Sox, and A. Tversky
1982   On the elicitation of preferences for alternative therapies. *New England Journal of Medicine* 306:1259–1262.

Milne, F.
    1981   Induced preferences and the theory of the consumer. *Journal of Economic Theory* 24:205–217.
Mishan, E.
    1976   Choices involving risk: Simple steps toward an ordinalist analysis. *Economic Journal* 86:759–777.
Morrison, D.
    1967   On the consistency of preferences in Allais' paradox. *Behavioral Science* 12:373–383.
Moskowitz, H.
    1974   Effects of problem representation and feedback on rational behavior in Allais and Morlat-type problems. *Decision Sciences* 5:225–242.
Mowen, J., and J. Gentry
    1980   Investigation of the preference-reversal phenomenon in a new product introduction task. *Journal of Applied Psychology* 65:715–722.
Pareto, V.
    1909   *Manuel d'Economie Politique*. Paris: V. Girard et E. Brière.
Parkin, M., and A. Nobay, eds.
    1975   *Current Economic Problems*. Cambridge: Cambridge University Press.
Payne, J., and M. Braunstein
    1971   Preferences among gambles with equal underlying distributions. *Journal of Experimental Psychology* 87:13–18.
Payne, J., D. Laughhunn, and R. Crum
    1980   Translation of gambles and aspiration level effects in risky choice behavior. *Management Science* 26:1039–1060.
    1981   Further tests of aspiration level effects in risky choice behavior. *Management Science* 27:953–958.
Pommerehne, W., F. Schneider, and P. Zweifel
    1982   Economic theory of choice and the preference reversal phenomenon: A reexamination. *American Economic Review* 72:569–574.
Pratt, J.
    1964   Risk aversion in the small and in the large. *Econometrica* 32:122–136.
Quiggin, J.
    1982   A theory of anticipated utility. *Journal of Economic Behavior and Organization* 3:323–343.
Raiffa, H.
    1961   Risk, ambiguity, and the Savage axioms. *Quarterly Journal of Economics* 75:690–694.
    1968   *Decision Analysis: Introductory Lectures on Choice Under Uncertainty*. Reading, Mass.: Addison-Wesley.
Reilly, R.
    1982   Preference reversal: Further evidence and some suggested modifications of experimental design. *American Economic Review* 72:576–584.
Roberts, H.
    1963   Risk, ambiguity, and the Savage axioms: Comment. *Quarterly Journal of Economics* 77:327–336.
Röell, A.
    1987   Risk aversion in Quiggin and Yaari's rank-order model of choice under uncertainty. *Economic Journal* 97(Suppl.):143–159.

Ross, S.
  1981   Some stronger measures of risk aversion in the small and in the large with
          applications. *Econometrica* 49:621–638.
Rothschild, M., and J. Stiglitz
  1970   Increasing risk: I. A definition. *Journal of Economic Theory* 2:225–243.
  1971   Increasing risk: II. Its economic consequences. *Journal of Economic Theory*
          3:66–84.
Russo, J.
  1977   The value of unit price information. *Journal of Marketing Research* 14:193–201.
Russo, J. G. Krieser, and S. Miyashita
  1975   An effective display of unit price information. *Journal of Marketing* 39:11–19.
Samuelson, P.
  1947   *Foundations of Economic Analysis.* Cambridge, Mass.: Harvard University Press.
          Enlarged edition, 1983.
  1952   Probability, utility, and the independence axiom. *Econometrica* 20:670–678.
  1977   St. Petersburg paradoxes: Defanged, dissected, and historically described.
          Journal of Economic Literature 15:24–55.
Savage, L.
  1954   *The Foundations of Statistics.* New York: John Wiley and Sons. Revised and
          enlarged edition published by Dover Publications (New York, 1972).
Schlaifer, R.
  1969   *Analysis of Decisions Under Uncertainty.* New York: McGraw-Hill Publishing
          Co..
Schmeidler, D.
  1989   Subjective probability and expected utility without additivity. *Econometrica*
          57:571–587.
Schoemaker, P.
  1980   *Experiments on Decisions Under Risk: The Expected Utility Hypothesis.* Boston:
          Martinus Nijhoff Publishing.
Schoemaker, P., and H. Kunreuther
  1979   An experimental study of insurance decisions. *Journal of Risk and Insurance*
          46:603–618.
Segal, U.
  1984   Nonlinear Decision Weights with the Independence Axiom. Unpublished man-
          uscript, University of California, Los Angeles.
  1987   The Ellsberg paradox and risk aversion: An anticipated utility approach.
          International Economic Review 28:175–202.
Shafer. W.
  1974   The nontransitive consumer. *Econometrica* 42:913–919.
  1976   Equilibrium in economies without ordered preferences or free disposal. *Journal
          of Mathematical Economics* 3:135–137.
Sherman, R.
  1974   The psychological difference between ambiguity and risk. *Quarterly Journal of
          Economics* 88:166–169.
Sinn, H.-W.
  1980   A rehabilitation of the principle of sufficient reason. *Quarterly Journal of
          Economics* 94:493–506.
Slovic, P.
  1969a   Differential effects of real versus hypothetical payoffs on choices among gambles.
          Journal of Experimental Psychology 80:434–437.

1969b Manipulating the attractiveness of a gamble without changing its expected value. *Journal of Experimental Psychology* 79:139–145.

1975 Choice between equally valued alternatives. *Journal of Experimental Psychology: Human Perception and Performance* 1:280–287.

Slovic, P., and S. Lichtenstein

1968 Relative important of probabilities and payoffs in risk taking. *Journal of Experimental Psychology* 78(3, Part 2): 1–18.

Slovic, P., and S. Lichtenstein

1971 Comparison of Bayesian and regression approaches to the study of information processing in judgment. *Organizational Behavior and Human Performance* 6:649–744.

Slovic, P., and S. Lichtenstein

1983 Preference reversals: A broader perspective. *American Economic Review* 73:596–605.

Slovic, P., and A. Tversky

1974 Who accepts Savage's axiom? *Behavioral Science* 19:368–373.

Slovic, P., B. Fischhoff, and S. Lichtenstein

1982 Response mode, framing, and information processing effects in risk assessment. Pp. 21–36 in R. Hogarth, ed., *New Directions for Methodology of Social and Behavioral Science: Question Framing and Response Consistency*. San Francisco: Jossey-Bass.

Smith, V.

1969 Measuring nonmonetary utilities in uncertain choices: The Ellsberg urn. *Quarterly Journal of Economics* 88:324–329.

Sonnenschein, H.

1971 Demand theory without transitive preferences, with applications to the theory of competitive equilibrium. Pp. 215–233 in J. Chipman, L. Hurwicz, M. Richter, and H. Sonnenschein, eds., *Preferences, Utility, and Demand*. New York: Harcourt Brace Jovanovich, Inc.

Spence, M., and R. Zeckhauser

1971 Insurance, information, and individual action. *American Economic Review* 61:380–387.

Stigler, G., and K. Boulding, eds.

1952 *Readings in Price Theory*. Chicago: Richard D. Irwin.

Stiglitz, J.

1975 Information and economic analysis. Pp. 27–52 in M. Parkin and A. Nobay, eds., *Current Economic Problems*. Cambridge: Cambridge University Press.

1985 Information and economic analysis: A perspective. *Economic Journal* 95(Suppl.): 21-41.

Stiglitz, J., ed.

1966 *Collected Scientific Papers of Paul A. Samuelson*, vol. 1. Cambridge, Mass.: MIT Press.

Sugden, R.

1986 New developments in the theory of choice under uncertainty. *Bulletin of Economic Research* 38:1–24.

Thaler, R.

1980 Toward a positive theory of consumer choice. *Journal of Economic Behavior and Organization* 1:39–60.

1985 Mental accounting and consumer choice. *Marketing Science* 4:199–214.

Tversky, A.

1969 Intransitivity of preferences. *Psychological Review* 76:31–48.

1975   A critique of expected utility theory: Descriptive and normative considerations. *Erkenntnis* 9:163–173.

Tversky, A., and D. Kahneman
    1971   Belief in the law of small numbers. *Psychological Bulletin* 2:105–110.
    1974   Judgment under uncertainty: Heuristics and biases. *Science* 185:1124–1131.
    1981   The framing of decisions and the psychology of choice. *Science* 211:453–458.
    1983   Extensional vs. intuitive reasoning: The conjunction fallacy in probability judgment. *Psychological Review* 90:293–315.

Viscusi, W.
    1985a   Are individuals Bayesian decision makers? *American Economic Review Papers and Proceedings* 75:381–385.
    1985b   A Bayesian perspective on biases in risk perception. *Economics Letters* 17:59–62.

von Neumann, J., and O. Morgenstern
    1947   *Theory of Games and Economic Behavior*, 2d ed. Princeton, N.J.: Princeton University Press.

Weber, M., and C. Camerer
    1987   Recent developments in modeling preferences under risk. *OR Spektrum* 9:129–151.

Weinstein, A.
    1968   Individual preference intransitivity. *Southern Economic Journal* 34:335–343.

Williams, C.
    1966   Attitudes toward speculative risks as an indicator of attitudes toward pure risks. *Journal of Risk and Insurance* 33:577–586.

Yaari, M.
    1969   Some remarks on measures of risk aversion and on their uses. *Journal of Economic Theory* 1:315–329.
    1987   The dual theory of choice under risk. *Econometrica* 55:95–115.

8
Conclusions

In valuing risk reduction and human life extension, the substantive regulatory problems encountered by benefit-cost analysts are often significant but are far from standardized. They vary in terms of the legal and administrative context in which they appear, the types of potential threats they pose to life and health, and the characteristics and availability of information about those threats. The family of approaches and techniques we have labeled *benefit-cost analysis* has not been systematically evaluated for its application to specific topical and conceptual problems in environmental health and safety. Moreover, as practiced in regulatory contexts, benefit-cost analysis is limited by the complex administrative, legal, and political process that characterizes health and safety policy making. The steering commmitee, on the basis of its conference papers and discussions, believes that the current challenge for those who perform or use benefit-cost analysis is twofold:

- to acknowledge and adequately identify the limited role of benefit-cost analysis and analysts in the entire process of regulatory policy formation and enforcement and
- to distinguish systematically applications of current risk-control approaches and techniques that are appropriate for specific types of regulatory issues.

These themes run through the comments of many conference participants, including those with sharply opposing views on specific topics debated in the various sessions. Several scholars, for example, point out that benefit-cost analysis is a small part of a very large, continuous decision-making process for health and safety regulation that also encompasses Congress, many administrative agencies and interest groups, scientists, the courts, and public perceptions. Some question the ability of those charged with valuation analyses to convey adequately uncertainties associated with the underlying scientific information on the risks of pollutants and toxins.

Others focus on the need for decision makers and analysts to consider a broader range of factors, including the concerns of future generations, in current analyses and decisions. Still others express a disbelief in the ability of any analysis that depends on trade-offs at the margin to reach what are considered by them to be absolute goals, such as health and environmental protection. Some practicing benefit-cost analysts wonder if they are asked to do too much with a set of tools that does not yet enable them to meet those expectations.

Taking into account the range of perspectives present at the conference, the steering committee noted several problems related directly to the role of analysis and analysts and the need to distinguish appropriate applications. For example, there is serious disagreement among scholars, analysts, and others familiar with the use of benefit-cost analysis as to whether current approaches to characterizing and valuing risks can or should be asked to accommodate the full range of factors that decision makers must implicitly take into account, especially for those problems involving intertemporal or purely qualitative comparisons. Even those who would support what they believe to be appropriate use of benefit-cost analysis in environmental policy making express the view that current methods can be employed more effectively in evaluating some types of issues than others—for example, employment effects compared with quality-of-life effects. In particular, there may be an irreducible tension between a desire to reduce value considerations to a single metric and a desire for symbolic protective action in health and safety regulation.

A related problem is the one of expectations versus constraints in doing benefit-cost analysis for regulatory policy making. For any area of regulation, legal, administrative, and political factors can set severe constraints on the conduct of analysis, as well as on the weight given to analytic results. In contrast, there are incentives, some in response to these constraints, for analysis and analysts to broaden their scope to show that all relevant factors have been treated.

Consequently, benefit-cost analysis cannot now be considered to be a formal decision-making mechanism accounting, for example, for the need for symbolic action or the full range of qualitative costs and benefits associated with policy alternatives. Benefit-cost analysis is more appropriately used in conjunction with other factors as a set of information-gathering and organizing tools that may be used to support both decision making and the presentation of information to the public.

Therefore, we believe there is a need to develop commonsense criteria for applying current methods to problems encountered in health and safety regulation, by defining in terms helpful to both analysts and critics alike the limits to analysis and by developing systematic application of analytic approaches and techniques to appropriately matched policy issues. Of great interest to analysts and decision makers is the practical impact of these issues on their work. In particular, conference participants returned repeatedly to the implications of considering benefit-cost analysis as a set of approaches intended to clarify and simplify certain parts of difficult problems raised in the course of environmental, health, and safety regulation. Much discussion focused on the current and potential roles for agency staff analysts in increasing understanding among decision makers and the public of the costs and benefits of regulation, including dilemmas and trade-offs inherent in real decisions.

Expression of these concerns, as well as suggestions for dealing with them, fall largely into five categories reflecting the most important aspects of the analytical process for environmental and other regulatory decisions:

- the administrative and legal context in which analysis is done, including its purpose;
- the overall approach that is or should be taken in benefit-cost analysis;
- issues related to the specific analytic procedures and techniques;
- the adequacy of underlying scientific information on risk (e.g., dose response or exposure); and
- the way in which decision makers incorporate risk assessments and analyses intended to clarify or support policy making.

Conference participants were asked to identify opportunities for improving risk-control analyses by practical suggestions for analysts and decision makers at EPA and other regulatory agencies charged with integrating the full range of intangible factors bearing on health and safety regulation. In considering this question, they were asked to define the limits of formal risk-control approaches and techniques.

Participants were not expected to nor did they reach consensus on any recommendation for a single set of analytical procedures or on a single appropriate role for analysis and analysts in the regulatory policy process. The cross-cutting concern for most participants remains the improvement of benefit-cost analysis as a practical set of tools for policy making. We believe that improving analysis involves careful consideration of underlying assumptions, concepts, and methods. It also involves assignment of appropriate types and levels of responsibilities to analysts (and to decision makers) in light of the analytical capabilities available at present.

THE CONTEXT OF DECISION MAKING

No actions within organizations take place in a vacuum—the context limits the possibilities or even favors the opportunities available to the analyst or the decision maker. These forces originate both outside the organization (e.g., in statutory pre- or proscriptions) and inside (e.g., in policy priorities). In terms of environmental decisions by the federal government, the interplay between the departments or agencies and the Congress constitutes one important set of external constraints. Statutes and legislative history, which vary considerably between agencies (and sometimes between programs within agencies) in terms of their approach to benefit-cost analysis and other valuation techniques, constitute another set of constraints. The reaction of the courts to the uses of valuation approaches in support of environmental regulation decisions is another, the activities of environmental interest groups and other organizations represent still another, and the particular policy goals of agency leaders are another. In his paper on the contextual setting for benefit-cost analysis, Melnick details the sources of several significant constraints and their major effects on the conduct and use of benefit-cost analysis for environmental decisions. In particular, the adversarial tone associated with policy making for much of health and safety regulation has the effect of providing extremely strong, but rarely singular, messages to benefit-cost analysts regarding how analysis should be done and used.

Statutes, court decisions, congressional intentions, and executive branch directives provide multiple, sometimes conflicting, guidance to agency decision makers and analysts on whether or how to include benefit-cost considerations in risk-management activities. Many key environmental health statutes operate with admonishments to "protect health" or "use best available control technology (BACT)," rather than offering a calculus that, where appropriate, includes trade-offs and cost-based alternatives. Taken as a whole, judicial rulings and opinions have likewise exhibited inconsistency across agencies, statutes, and even specific policy arenas. Congressional attitudes are often negative toward valuation approaches, while OMB has pressured decision makers and analysts to more fully integrate such approaches into risk management.

Regulatory analysis reflects these characteristics, in that many of its underlying assumptions are constrained by these legal, administrative, and political features. For example, an analysis of fugitive arsenic emissions from copper smelting would vary considerably depending on whether it is deemed important to preserve the industry in question (see the appendix, which takes up this question). It is clear that few expect federal health, safety, and environmental regulation to become more consistent and consensual, either legally, bureaucratically, or politically, without major legislative intervention. The most compelling questions concern prospects for major intervention and how environmental analysis might be done in light of these characteristics. For example:

- How can benefit-cost analysis be integrated into statute-based regulatory management practices that present a wide range of stances toward analysis (e.g., BACT versus health-based standards)?
- Can we expect benefit-cost analysis to become more politically feasible or acceptable?

In response to these questions, there were two types of action-oriented approaches expressed at the conference, one focusing on changing the environment for benefit-cost analysis and the other focusing on improving analysis in response to key criticisms of its assumptions and techniques. In particular, participants discussed the need and prospects for (1) creating a more consistent and conducive context for valuation approaches by amending the key environmental health statutes and (2) modifying analytical techniques and/or including additional variables in valuation approaches in order to generate support for the results of benefit-cost analysis among the public agencies, the courts, and Congress.

Achieving improved statutory and judicial consistency and encouragement for the use of valuation approaches in health and safety decisions would most likely rest on three things. First, general agreement among policy makers would be needed that there is a compelling need to compare gains in environmental health with the cost of achieving those gains. Second, agreement would be needed that previous improvements in the techniques of benefit-cost analysis now enable sophisticated assessments of variables, such as human life extension, temporal effects, and other important issues. Third, acknowledgment would be needed that benefit-cost analysis is or can be decoupled from political motivations, such as the perception that valuation approaches may support or justify policies favoring industry or other special groups to the detriment of the wider public. However, agreement on the need for cost reduction and/or greater efficiency in protective regulation clashes with concerns that efficiency can be used inappropriately as a rhetorical device justifying efforts to remove or reduce needed regulations regardless of their allocative implications.

While benefit-cost analysis cannot now be considered to be a single set of valuation approaches, techniques, and procedures, there is a core of common understanding among many analysts based on shared assumptions regarding expected utility, the implications of scarce resources, and other concepts. Indeed, explicit techniques for valuing risks, costs, and benefits have existed for several decades. However, one effect of multiple, conflicting guidance from the regulatory context—even for a single agency—is that regulatory requirements and practices for the use of valuation approaches are accompanied by political and policy forces favoring or opposing their application. Consequently, the burden to develop criteria matching appropriate approaches and valuation techniques to specific types of problems may be too large. For example, many believe that the use of benefit-cost analysis is not consistent with the Clean Air Act (benefits can be examined but not costs) and the Clean Water Act (analysis of costs but not benefits is allowed), and yet benefit-risk analyses are required by the Federal Insecticide, Fungicide and Rodenticide Act.

There are at least two different sets of perspectives on these questions. One view is that the crux of the battle over acceptability of benefit-cost analysis is a debate about how much power should be given to those who favor and perform such analyses. According to this view, the battle rages between Congress and the regulatory agencies, as well as between groups residing within the regulatory agencies. Another view reflecting genuine puzzlement in the case of environmental trade-offs sees polarization and controversy emerging from institutions (e.g., Congress, the bureaucracy) that are considered on most subjects to be moderating and consensus building.

For both views, it may be that the use of benefit-cost analysis for environmental health touches on basic unresolved differences regarding the appropriateness of applying cost considerations to certain classes of public goods, such as those described in the papers by Railton and MacLean. If so, these issues must be addressed at the level of institutions (i.e., Congress, the President, and the courts), and possibly through the conduct of further research, before specific guidance can be supplied to regulatory decision makers and analysts. Such a resolution might emerge over the long term or as the result of a major environmental or economic crisis. But, in the absence of a crisis, public opinion and institutions might continue to have difficulty integrating fully the protection of human health and safety and the consideration of costs and benefits in allocating scarce economic and other resources. Signals from the legal and administrative context, furthermore, suggest that agencies may continue to find both strong supporters and strong critics of attempts to enunciate agency-specific or government-wide approaches to risk-control analysis for regulation.

On the second question—whether future improvements in analytic techniques, approaches, and theories could prove compelling to the policy-making system and therefore generate additional support for the consistent use of valuation approaches—support for (or opposition to) benefit-cost analysis may not always be based on assessments of technical features. Therefore the contextual forces for regulatory benefit-cost analysis may not respond to technical improvements with greater support for its use in regulatory decision making.

Nevertheless, it was clear to many participants that any advances in the application of evaluation approaches must be grounded solidly in the quality of the analysis and its ability to speak to values considered important by citizens and policy makers, regardless of whether they are easily discussed analytically. Consequently, much of the conference focused on ways to account for these additional variables and questions in valuation approaches used in benefit-cost analysis.

APPROACH

The term *approach* refers here to the way issues are conceptualized in terms of a general framework for analysis. Benefit-cost analysts, as Melnick notes, typically see policy issues in terms of opportunity costs, incentives, and the expense associated with eliminating risks. He characterizes lawyers, in contrast, as more likely to give preference to concepts of "command and control," of establishing standards and penalties to be imposed if those standards are not met. At least some environmentalists and politicians, he says, are more likely to see the issues in terms of an absolutist, health-only stance for which no level of effort is too great. To the extent that these characterizations are accurate, they influence the way analytic studies of valuation are viewed.

In order to agree on an approach to valuing health effects, we would want it to pass three tests: (1) accuracy, (2) verifiability, and (3) accept-ability or feasibility. Much of the technical work to advance health-effect valuations has focused on improving the accuracy and verifiability of the scientific data and benefit-cost techniques. MacLean, however, describes some of the reasons why this third test presents special difficulties. He points out that there are substantial differences in the positions favored by different groups, differences that are difficult to resolve through appeals to either basic moral principles or empirical preferences. MacLean describes what for many makes the issue absolutely unique: human life is sacred, a fact with many ramifications. For example, although few disagree that it makes sense to discount expenditures or the opportunity costs of health effects, MacLean claims that there are moral and logical difficulties when the value of life per se is discounted. He argues that the existence of eco-nomic consequences of health and safety decisions does not justify treating human life as if it were exchangeable.

There remain unavoidable comparisons, however, of how many re-sources should be devoted to improved physical well-being compared with energy conservation, greater economic growth, or other outcomes. One view is that such comparisons can be made more efficiently by relying on a common metric to characterize all important possibilities rather than examining the relative value of risk reduction in each case. Yet even within the community of professional analysts, there is disagreement about the best way of doing so. For example:

- Where are improvements most needed in valuation approaches?

- Are there viable alternatives to formal valuation approaches to support environmental decisions?

As the above discussion of contextual issues indicates, a major challenge for analysts is to develop and employ appropriate approaches to analysis when there is disagreement regarding basic policy problems and goals. For example, some have proposed that air and water pollution could be reduced more rapidly by making standards less stringent. The reasoning is that, in some cases, current standards so far exceed current attainment levels that there is little chance additional efforts will lead to complete compliance. Therefore, the argument goes, fewer such efforts are undertaken than would be if standards were set at a lower level. This reasoning is rejected by those who believe that high standards should be met. Still others would acknowledge the difficulty of meeting some current standards, but hold that they act as a necessary spur to obtaining even the most modest pollution reduction results. In this context, how does the analyst choose between methods of analysis appropriate to obtaining relative reductions or to meeting the more absolute goals?

In addition, some analysts and observers are concerned about the accessibility of analytic approaches. Many accept that an important attribute of an ideal approach should be that it generate results people can understand and can analyze and argue about. In this sense, Railton refers to benefit-cost analysis as an information-yielding device rather than a decision-making tool. There remains, of course, the question of accessibility and training. Results easily understood by another benefit-cost analyst may not be fathomable by an interested, educated, but untrained person or even by environmental scientists trained in other disciplines.

Machina presents what might be considered by some to be an unusual position with respect to the various parts of the analytical process that need adjustment. He argues, for example, that recent theoretical and empirical findings pertaining to how people actually evaluate risks and assign probabilities ought to be reflected in the way preferences are analyzed, but not in the analysis of probable outcomes. As he states it, observed "departures from the strictures of probability theory should be corrected but . . . (systematic) departures from the strictures of expected utility theory should not". This is because the former involve the determination of the risks associated with alternative actions or policies, which are in fact matters of accurate representation, while the latter involve the willingness of individuals, organizations, and society to bear these risks, which is a matter of preference. He concludes that analysis must be designed to account for actual preferences, even those that depart from the strictures of expected utility theory.

Considerable support for this view is emerging among economists and other risk-control analysts. However, there appears to be little or no consensus at present regarding specific valuation approaches that need reforming in light of this new evidence on preference formation and little consensus as to how such approaches, once identified, might be modified to systematically account for observed departures from expected utility theory.

Furthermore, there is little agreement as to alternatives to formal valuation approaches in support of environmental decisions. Although many decry weaknesses in current analytic approaches, few alternatives are proposed. MacLean discusses several reasons that a single analytic method for making environmental decisions should not be applied universally. This is principally because the necessity to involve symbolic elements in comparisons and trade-offs makes such decisions very dependent on their context. He concludes that it may be necessary to treat different values differently and that situational specificity, not measurement difficulty, is the fundamental problem in such comparisons. As Railton points out, an approach that yields more information can increase accuracy yet at the same time be less decisive.

Regardless of the valuation approach used, most participants seemed to favor greater provision of information, even at the cost of decreased decisiveness. Still, there is also a recognition of the need to systematically differentiate analytical approaches in light of the problems or values they are meant to address.

PROCEDURE

Procedure, as used here, refers to the way the benefit-cost analyst actually goes about systematically examining values. The choice of specific methods, metrics, and measures will play an important role in providing understandable, believable information for decision makers and other interested parties in protective regulation. For example, the concept of willingness to pay (i.e., setting the value of a good or service according to what people are willing to forgo in order to have that good or service) has gained adherents and wider use in health and safety regulation in recent years. When used in hypothetical markets (e.g., environmental decision making), it is nevertheless a troublesome procedure to those who would prefer to incorporate the notion of experience in some way, and who would attempt to discover what people, *with enough experience*, would want. Another possibility would be to balance willingness to pay with a measure of willingness to sell (how much people would have to receive in order to give up the good or service), but such a concept also runs into measurement problems in hypothetical markets.

Questions are raised also about the advisability of relying on tort law to determine the value of intangibles, especially because jury awards appear to be highly skewed in liability cases. Another possibility is to search for valuation methods and/or procedures that account for distributional effects, for example, assigning premiums to consequences based on income. Other questions include:

- Can balanced procedures, possibly including a combination of valuation techniques and nonquantitative factors, be found that would meet the approval of Congress, OMB, and the courts?
- Is the use of a simple or single metric appropriate when there is variation among individuals, not just in terms of how much of a value is wanted or can be tolerated, but in terms of whether they want that particular value at all?
- Are there ways to capture certain difficult conceptual problems, such as intergenerational equity, in analytic terms?

In his keynote address in this volume, Russell describes the necessity of making what he calls "cruel choices." In terms of the accommodation sometimes needed to reach agreement about such choices, it may seem as though valuation techniques make trade-offs too explicit. Fuzziness sometimes appears to have a positive function in that it allows the participants to believe in the myth of successful joint accommodation. Russell argues, however, that better information needs to be presented to the public, and that our system is predicated on the legitimacy flowing from the support of an informed citizenry. Presenting these trade-offs is a critical part of that process.

One procedure for simplifying, analyzing, and presenting trade-offs is the transformation of a range of value observations into a single set of common metrics. In analytic practice, this most often means the use of monetary value as the common numerator or denominator, or both. Railton discusses the difficulty of relying on an expert-determined metric when issues of welfare are involved. Scientists, he says, have more knowledge about natural phenomena and thus would be more reliable than the rest of us in describing events. But the most reliable source for describing people's welfare outcomes may be those individuals themselves. It may be quite difficult to develop a metric that adequately captures individuals' perceptions of their well-being. Nevertheless, in practice the regulatory process does implicitly differentiate certain factors. Some have observed that regulators are willing for society to expend more to protect individuals who are involuntarily at risk than to reduce those risks borne by people who are easily able to make choices regarding their own exposure. This corresponds, by the way, to many people's preferences concerning a variety of risks.

Similarly, the problem of comparing well-being across a significant span of time (e.g., intergenerational comparisons of health) has provoked considerable controversy about the acceptability of analytic procedures. There are regulatory actions involving human health and life for which there is a need to consider costs and benefits that might accrue in the future. The traditional method for comparing costs and benefits over time—discounting—is predicated on the belief that costs borne by and benefits accruing to individuals and society in the future should be adjusted, or discounted, to make comparison with current expenditures or costs valid. In the area of human health and safety, the sacredness of human life, as MacLean and others characterize it, can make it difficult for policy makers to agree with methods that seem to say that a life saved today should be worth more than a life saved 50 years from now. On the other side of the coin, if resources are limited, then some allocation must occur. Under those circumstances, the hard question is: What is the best way to deal with policy problems for which actions taken or not taken today will have human health effects in the distant future?

In answer to these difficulties and questions, there is at present no broad consensus among those interested in health and safety regulation. Some oppose discounting altogether, others would apply discounting to costs but not benefits, and many would support the use of a very small discount rate far below that often used in analyzing costs and benefits of other government and private programs.

It is important to note that although (as previously discussed) benefit-cost analysis may be only a small component of the entire regulatory process, analytic procedures must be constructed to account for the major characteristics of decision making. Harris raises the point that, whereas the usual way of thinking about analysis is as an input to decision making, decision making often effectively provides input for subsequent analysis, and that the best strategy may be to treat the whole process as dynamic. The factors that ought to be considered may be too complex to model adequately using current analytic methods, and the only way of developing appropriate information about outcomes may be to take action that can be reversed if necessary.

SCIENTIFIC INFORMATION

The potential conflict between risk assessment and risk-control analysis is often realized. In the schema of key components of regulatory agency policy making presented in the introduction to this volume (see Figure 1), risk assessments depend on assessments of data from epidemiologic, biological, and engineering studies. In turn, benefit-cost analysis depends on the application of valuation analyses to scientific assessments of risks associated with health and safety problems. Both risk assessment and risk-control analyses inevitably involve simplification of processes and outcomes in order to achieve clarity, meet regulatory deadlines, or because the underlying data are insufficient. For example, scientists frequently complain that risk-control analyses fail to capture adequately the factors they judge important in scientific determinations, whereas analysts express disappointment at being prevented from achieving the ideal in risk assessment, accurately and precisely portraying the risks of concern to policy makers and the public. This, both claim, is because some factors can be quantified with confidence while others remain matters of judgment. Scientists, if generalization may be permitted, tend to distrust the policy analysts as oversimplifiers, and the analysts are frustrated because the scientists cannot provide clear-cut answers. Several practical questions are embedded in this caricature:

- Is it possible to adequately reflect scientific judgment in analytic valuation studies?
- Can uncertainties or gaps in data be adequately incorporated?
- Can conflicting models or interpretations be integrated into analytic valuation studies?
- How can values or policy judgments be integrated into analytic valuation studies?

The ways different disciplines treat data, uncertainty, and judgment may underlie these issues. For example, some natural scientists point out that their colleagues are currently unable to come up with a method for quantifying, for example, the carcinogenicity of arsenic that would satisfy analysts interested in determining the costs and benefits of policy options for dealing with fugitive emissions associated with copper smelting. These scientists were greatly disturbed that an enormous body of information, mechanistic as well as epidemiologic, about the neurotoxicologic effects of arsenic was ignored in the analysis of the substance as a human carcinogen because it is judgmental and therefore difficult to express in quantitative terms.

In his paper on information and regulation, Harris raises a related issue having to do with the timing and sequence of new scientific information as it becomes available for benefit-cost analysis. The question is knowing when to act and when to wait for additional information that might lead to a different action (including no action). The challenge shared by the scientist, the analyst, and the decision maker is to construct procedures for knowing when to recommend action, further information gathering, or both.

Scientists, analysts, and decision makers, however, point out a number of other issues that follow from an expansion of the concept of risk assessment to include other factors. For example, some emphasize the distinction between narrow technical uncertainty associated with any scientist's estimate of risk and the broader structural uncertainty associated with differing estimates of risk by various scientists. Most important, others argue that a major reason why better scientific estimates of health and safety risks are not available—for example, estimates of the risks of noncarcinogens—is that decision makers and the public have not demanded them. Estimates of the risks of carcinogens are relatively more available largely because Congress, the courts, and the public have demanded that resources be committed to producing those estimates.

Finally, there is a significant argument that risk information is often inadequate because scientists are reluctant to make estimates in light of limited data. This may be a major contributing factor in the conflict between analysis and scientific risk assessment. Benefit-cost analysis is often driven by the need to make a decision, while those who perform scientific risk assessments are reluctant to guess when information is deemed inadequate.

It may be that scientists will always be distrustful of benefit-cost analyses, which, in order to reduce, simplify, and meet tight schedules, rely on summary characterizations rather than detailed information and accept increased level of uncertainty and a reduced level of subtlety in understanding any single piece of the whole picture. There is a conflict, however, between presenting complete information that is so detailed and complex that only the experts can understand it, and simplifying presentations so that those without the technical training can understand but lose the detail and subtlety of the data. There does not appear to be a quick technical or procedural fix that satisfies both scientific sophistication and simplicity. But many would insist on making evidence, and especially the assumptions underlying conclusions, regarding both risks and the valuation of risks as open and explicit as possible. This would both encourage careful questioning of those assumptions and contribute to trust among scientists, analysts, decision makers, and the public.

DECISION MAKING

A widely shared expectation is that an individual or small group, usually at the apex of the regulatory agency, holds responsibility for the decisions and actions of that agency. Another shared expectation is that a policy decision should be based on the relevant evidence. As indicated previously in this discussion, the role of the agency analyst, as well as that of the agency decision maker(s) at the apex, is often limited by a context in which myriad other voices, both outside and inside the agency, speak strongly. Moreover, these two expectations can conflict, particularly when it is difficult to gain general agreement on the type and quality of evidence needed to support a decision or when some important evidence is qualitative, implicit rather than explicit, or missing altogether.

In this type of setting, agency political appointees, rather than analysts, are responsible for factoring in a full range of considerations bearing on a problem—in effect, to act on behalf of the agency and the public by applying an informal metric trading off the economic, political, ethical, and etiological concerns raised in the foregoing papers—and then to produce a decision reflecting the results.

Within this framework, benefit-cost analysis can be used to provide more explicit information about some (e.g., economic) priorities and trade-offs as input to a decision, or it can be used to justify decisions already reached. It is not intended, however, nor can it give equal emphasis to all possible considerations bearing on a regulatory decision. Rather, analysis traditionally focuses on a few factors the decision maker would like to make explicit or those few for which quantitative values are available.

- What are the prerequisites for analysis that informs rather than appears to supplant the prerogative of the decision maker?

There is some ambiguity regarding how much authority we wish to grant to the process of providing information and the experts who provide it. Both analysts and policy makers have experienced pressure to use benefit-cost analysis to treat explicitly a broader range of relevant factors. This includes direct statutory and administrative pressures for agencies to broaden or to restrict their analytical coverage of values, including economic, etiological, ethical, and administrative concerns. In addition, strong criticisms and attacks on risk-control analysis by those mistrustful of the assumptions, approaches, and results of valuation approaches may play a role in stimulating theorists and practitioners to increasingly incorporate and analyze as many issues and aspects bearing on a decision as possible.

The analyst, consequently, faces incentives to include a wider range of considerations within the analytical framework, considerations that may be difficult to treat formally (e.g., ethical and political variables). However, as the number of variables to be addressed increases and as their character changes (e.g., inclusion of variables that are intangible or those that raise questions of interpersonal comparisons), the analyst's challenge is to prevent the problem under consideration, as well as the set of alternative decisions, from becoming fuzzier and less well defined. Some worry that, as the analyst incorporates an increasing number of factors, including those factors not often treated quantitatively, a formal but limited set of approaches and techniques is being substituted for the more informal approach provided by the politically and legally accountable decision maker.

Three principal prerequisites are often cited as expectations for analytic studies: knowledge, candor, and clarity. With regard to the first prerequisite, most studies need three kinds of knowledge: technical knowledge, economic knowledge, and legal knowledge. There is a lot of uncertainty in all three areas, and whatever is presented to the decision maker must reflect the degree of confidence in the knowledge in those three areas.

Second is candor. If the people doing the studies know that there are some weaknesses or driving assumptions embedded within them, they should frankly state that to the decision maker. At issue is the generation of trust among those who experience the effects of regulation, among the decision makers, and among valuation analysts. Much has been written in recent years about the decline in trust in government and other institutions. We need not repeat that discussion here, other than to point out that many factors affect trust and that it is important for analysts to demonstrate substantive competence, lack of bias, and openness in their work.

Finally, whatever is presented to the decision maker should be clear. Studies may be clear to the leading workers in the field and their peers, but not to decision makers.

The peer review system represents one possible way to promote accurate knowledge, candor, and clarity in the use of benefit-cost analysis in regulation. Review by knowledgable experts can help reveal inaccuracies, weaknesses, driving assumptions, and so on. But it may not be capable of ferreting out the technical and other jargon that confuses nonexperts. Conference participants pointed out that, although peer review is commonly used to examine the methods, data, and results of risk assessment, formal peer reviews of benefit-cost analysis are rarely undertaken within regulatory agencies.

Few would support applying a mechanical, strict benefit-cost rule to environmental regulation that is not tempered by judgment and consideration of other factors. The major role of benefit-cost analysis in environmental regulation should be as an organizing concept and as a way of helping the decision maker think about the factors that need to be looked at. A process involving peer review of benefit-cost analysis could assist decision makers by making assumptions clear and by noting the strengths and weaknesses of results.

Railton's suggestion that benefit-cost approaches be treated as information-yielding rather than decision-making devices has several implications for the way scientific data and judgment could be integrated into decision processes. Most important, it would suggest disaggregated information and error bands or ranges rather than point estimates. It also suggests the importance of presenting information about several alternatives. Harris, moreover, points out that the best strategy may be to undertake actions that are reversible before complete information is available, because implementing regulatory or other control actions may be the only way to obtain relevant information. Information generation may be an excellent role for benefit-cost analysis in such a dynamic process.

CONCLUSIONS

The conference was intended to explore a set of philosophical, political, informational, and administrative issues in the use of benefit-cost analysis for environmental decision making. This broad scope yielded an equally broad mix of ideas, approaches, and conclusions in both the papers and the conference discussions. The ideas discussed at the conference and presented in these papers may contain a few seeds that could bear fruit if they are nurtured and developed. Certainly they include several concepts and comments that speak directly to the practice of benefit-cost analysis.

Emerging from conference discussions is a strong sense that, at present, the challenge in improving the application of benefit-cost analysis is to design practical procedures and techniques that accommodate (1) considerable situational variation; (2) the fairly limited role played by formal risk-control analysis in the full process of identifying, regulating, and enforcing solutions to health and safety problems; and (3) the tendency for both critics and supporters of analysis to overemphasize its influence in the regulatory process.

The problems of health and safety regulation are far from standardized. Similarly, benefit-cost analysis is really a family of related techniques and approaches, only a few of which have been systematically evaluated for their application to specific topical and conceptual issues in health and safety regulation.

Consequently, the application of benefit-cost analysis must be supported by better systematic distinctions or situational conditions that reflect accurately key variations among regulatory responsibilities and problems facing health and safety decision makers. For example, can we systematically distinguish among situations involving potential loss of life, sickness and injury only, or both?

Once a set of systematic distinctions has been drawn, it would then be important to identify attributes of benefit-cost analysis techniques appropriate to those situational conditions. At the simplest level, for example, are there some problems for which the use of formal analytical techniques are not currently appropriate? At another level, is it appropriate to use discounting of future costs and benefits when human life is not at stake?

Finally, it is important to formulate processes for encouraging the development, within the various federal regulatory arenas, of agreement on systematic scientific distinctions, techniques, and approaches to valuation of risks, costs, and benefits.

Taking account of the fact that some conference participants generally support the use of formal analysis whereas others generally oppose its use, the committee was nevertheless able to reach the following conclusions.

1. *Among both those who would generally support the use of benefit-cost analysis and those who would oppose its current use, there is a common recognition of genuine moral and ethical dilemmas underlying evaluation of the costs and benefits of programs to regulate health and safety risks.*

a. *Current approaches to characterizing and valuing risks cannot accommodate with validity the full range of factors that decision makers are asked to take into account, particularly those drawing comparisons across time.* Statutes and administrative orders requiring major regulatory decisions to be based on least cost/most benefit analyses have, according to one view, stimulated a new emphasis on cost reductions in regulatory decision making. According to another view, such requirements fail to acknowledge that the general approach to benefit-cost analysis is not developed well enough to fully account for important factors, such as public opinion, quality of life, and other social, psychological, and institutional issues important in any program of regulation. Similarly, considerable disagreement exists about the ability of risk-control approaches to adequately consider future generations with current techniques for discounting. Supporters have proposed careful analysis of long-term impacts of regulatory policies and the use of nonmonetary measures, such as lives saved/deaths prevented, in calculating impacts. Critics are concerned that equity cannot be discerned and compared across time in ways that can be formally measured. This suggests a need for further development in the areas of temporal and nonquantitative factors as a contribution to improvement in benefit-cost analysis.

b. *Serious concerns exist regarding the appropriateness of formalizing approaches to issues such as intertemporal equity.* At the heart of this recognition is the belief that there is an irreducible tension between a desire to reduce value considerations in regulation to a single metric and a desire for symbolic action. This conflict emerges most clearly with respect to issues involving human life (preventing deaths/saving lives) in regulatory decision making. There is a concern that, even if benefit-cost analysis approaches and techniques could be developed to account systematically for a broader range of issues on a broader range of regulatory topics, it may be morally or ethically incorrect to place values such as human life next to other economic factors.

2. *Formal benefit-cost analysis of health and safety risks in regulation is at present only a limited and incomplete part of a large, complex analytic and decision-making process.* This process largely determines the weight to be given to benefit-cost analysis in policy making, and it sets limits on what can and cannot be accomplished with formal evaluation of risks, costs, and benefits. For example, the character of statutes, regulations, court decisions, and enforcement can all constrain the assumptions, methods, and data that can be used in a formal analysis. And, once completed, the scientific quality of the analysis is only one factor influencing reactions to the analytical results.

3. *There are often misperceptions regarding the influence of benefit-cost analysis in decision making.* Critics often express the need to curb the power of benefit-cost analysis in decision making, while supporters would increase its influence. There are at least two general reasons why this polarization and possible overemphasis may occur. First, the current adversarial character of regulation for health and safety issues encourages some to seek ways to challenge (or defend) outcomes on whatever grounds may prove successful. Under these circumstances, analysis can become a target or a pawn of advocates. Second, there may be standards and/or goals set for analysis that it cannot in all instances meet. In a policy process in which analysis seems to offer ways of treating difficult issues, there is a tendency to transform political disputes into scientific or analytical ones and then to expect the analytical process to substitute for the political process. The consequence is to expect benefit-cost analysis to be able to accommodate a wide range of important factors bearing on a problem and then to be critical when it is more difficult to treat formally some factors than others. The result can be to politicize benefit-cost analysis and polarize positions (e.g., those who would increase the use of analysis and those who would eliminate it in favor of absolute standards, such as the Delaney Amendment and the Clean Air Act).

4. *We believe that, among agency decision makers, the courts, Congress, and analysts, there is no consensus regarding the use of a specific set of analytical techniques for a specific purpose.* At present there are a range of techniques available for use in benefit-cost analysis of morbidity, mortality, and other health and safety effects. There is agreement among many practitioners and scholars regarding the importance of considering a wide range of costs and benefits of alternative policies; there are also, in many instances, a number of technical procedures and methods that analysts can use to estimate those costs and benefits. But there is little agreement on decision rules an agency might use in choosing one procedure or method over another in specific cases. In the analysis of programs to prevent deaths—the example is regulation to restrict fugitive arsenic emissions from copper ore smelting—an agency analyst might choose techniques based on

willingness-to-pay. It is well known that this approach emphasizes different types of costs than would other approaches. For example, techniques based on willingness-to-pay will emphasize expressed preferences, while those based on human capital models emphasize forgone earnings. Although there is considerable debate within some agencies and among some analysts as to the appropriateness of specific techniques for valuing lives, there is a need to evaluate both the types of regulatory problems and the attributes and implications of techniques for benefit-cost analysis, in order to develop a more systematic basis for deciding when to apply which technique.

Based on these conclusions, the steering committee makes three recommendations for improving the use of benefit-cost analysis in health and safety regulation:

1. *Benefit-cost analysis should be thought of as a set of information-gathering and organizing tools that can be used to support decision making rather than as a decision-making mechanism itself.* Treating the approach in this way has several implications for the way it is applied. It emphasizes disaggregated information and range rather than point estimates. It encourages development of information about several alternatives rather than a single policy. It also encourages viewing both analysis and decision making as dynamic rather than static processes.

2. *There is a need to more systematically match analytic methods and techniques to types of health and safety problems encountered in the regulatory process.* Researchers and analysts recognize that techniques associated with benefit-cost analysis emphasize quantifiable factors, especially those that can be characterized in monetary terms. In addition, benefit-cost analysis is most applicable to those health and safety problems or elements of problems in which externalities are relatively limited and opportunity costs are relatively clear. For example, the economic impacts of closing a plant may be easier to characterize succinctly than its effects on the quality of life in the surrounding community. This implies (a) the need to improve benefit-cost analysis so that a more complete range of impacts and factors can be considered systematically and (b) the need to identify systematically sets or types of problems according to their susceptibility to benefit-cost analysis. Therefore, blanket determination of whether a single analytic approach should be taken to all regulatory decisions does not seem appropriate at this time.

3. *Regulatory agencies should consider expanding the use of formal peer review mechanisms in the area of benefit-cost analysis for health and safety decisions.* The steering committee believes that benefit-cost analysis should be subject to systematic, consistent, formal peer review, which can be used to assess appropriateness of assumptions, techniques, and approaches; limitations of data and methods; and the formal or informal treatment

of moral and ethical concerns. Traditionally, the Office of Management and Budget has been asked to assess major regulatory benefit-cost analyses as part of its mission to encourage greater use and improved quality in analysis, but it is not appropriate to ask OMB to devote its limited resources to accomplish a full range of review that is necessary for a complicated valuation. OMB could be asked to respond to plans for expanding peer review for valuation analyses. Such an expansion would complement existing agency peer reviews for scientific risk assessment prior to agency decision making.

Appendix
Setting National Standards
for Inorganic Arsenic Emissions
from Primary Copper Smelters:
A Case Study

RALPH A. LUKEN

Section 112 of the Clean Air Act requires the Environmental Protection Agency (EPA) to establish emission standards for hazardous air pollutants that protect public health with an "ample margin of safety." In interpreting the language for the purposes of regulatory development, EPA does not consider the word "safety" to imply a total absence of risk. Many activities involve some risk but are not considered "unsafe." In EPA's view, standards under Section 112 should protect against significant public health risks.

In setting a Section 112 standard, EPA identifies sources of pollution that may pose significant risks, determines the current and planned levels of control at those sources, and assesses the health risks associated with those levels. If a source is judged to pose a significant risk, EPA selects a level of control that, in its judgment, reduces the health risks to the greatest extent that can reasonably be expected, after considering the uncertainties in the risk analysis, the residual risks that remain after the application of the pollution control technology, the costs of further control, and the societal and other environmental impacts of the regulation. This process is referred to as risk assessment and risk management.

Policy analysts and decision makers long have struggled with how best to apply economic methods and assumptions when analyzing and controlling risk. On the one hand, the limitations of economic assumptions and

Ralph "Skip" Luken is chief of the Economic Studies Branch, Economic and Regulatory Analysis Division, in EPA's Office of Policy, Planning, and Evaluation. The views expressed in this case study are those of the author and not necessarily those of EPA.

analysis have been widely discussed and have in fact somewhat restricted the application of this approach to such areas as risk management. On the other hand, the necessity of recognizing the risk choices associated with and the trade-offs of regulatory options argues for some role to be played by economics in analysis and decision making.

Most debate over the use of economic principles and methods in risk assessment and risk management eventually focuses on the underlying assumptions (usually implicit) that relate to rational behavior, ethics, public choice, and time preference. At this level, the issues concern notions of equity, social values, philosophical presuppositions, and the social contract between government and its citizenry.

This case study is intended as an opportunity for the reader to apply the concepts of risk assessment and risk management to arrive at a decision about what constitutes an adequate level of health protection from one source of hazardous air pollution. In this case, the hazardous air pollutant is inorganic arsenic, and the source is uncontrolled fugitive emissions from primary copper smelters that process copper ore containing arsenic as an impurity.

The primary copper smelting industry in the United States uses pyro-metallurgical processes to extract copper from sulfide copper ores that contain arsenic as an impurity. At the 15 primary copper smelters operating in the United States in 1983, the average arsenic content of copper ore ranged from 0.0004 to 4.0 weight percent. The average arsenic content of the ore was well below 0.5 weight percent at the majority of smelters; only the Tacoma smelter processed ore with more than 1 percent arsenic.

The 14 low-arsenic copper smelters are the subject of this case study. Arsenic emissions from these smelters total 730 megagrams per year. Process operations, which emit about 530 megagrams per year, are already controlled to the extent technically possible. Uncontrolled fugitive sources account for the remaining 200 megagrams of arsenic emissions per year. The regulatory question is as follows: *How many* primary copper smelters should be required to control fugitive inorganic arsenic emissions?

On June 5, 1980, EPA published a *Federal Register* notice (U.S. Environmental Protection Agency, 1980) listing inorganic arsenic as a hazardous air pollutant under Section 112 of the Clean Air Act. On July 11, 1983, EPA proposed standards for inorganic arsenic emissions from the 14 low-arsenic primary copper smelters as well as from the single high-arsenic copper smelter (U.S. Environmental Protection Agency, 1983c). On August 4, 1986, EPA issued a final standard for inorganic arsenic emissions from the 14 low-arsenic primary copper smelters and withheld further action on the high-arsenic copper smelter because it had ceased operation (U.S. Environmental Protection Agency, 1986b).

The reader is encouraged to make his or her own regulatory decision, given the same information that was available to EPA in mid-1986, before reviewing the actual EPA rule making for inorganic arsenic.

The next section presents background risk assessment information as a scientific estimate of health risk. The following section presents background risk management information composed primarily of engineering and economic descriptions of the consequences of controlling emissions at the 14 smelters. The final section highlights some of the important factors to consider in determining the level of pollution control that would protect the public against significant health risks with an ample margin of safety.

RISK ASSESSMENT: QUANTIFYING CANCER RISKS

The quantitative estimates of public cancer risk presented in this section are based on (a) EPA's linear nonthreshold model, which numerically relates the degree of exposure to airborne inorganic arsenic to the risk of getting lung cancer; and (b) EPA's Human Exposure Model, which expresses numerically the degree of public exposure to ambient air concentrations of inorganic arsenic from the 14 copper smelters. This section describes these models and the assumptions used to assess cancer risks and presents EPA's quantitative estimates of individual and population risks. It also discusses uncertainties in the risk characterization that should be considered before preparing estimates of individual and population risks.

Estimated Dose Response

The numerical constant that defines the exposure (dose)/risk (response) relationship in the linear nonthreshold model is called the unit risk factor. For an air pollutant, the unit risk factor is the excess cancer risk associated with an individual's lifetime of exposure (70 years) to an average concentration of 1 microgram per cubic meter (1 $\mu g/m^3$) of the pollutant in the air.

For inorganic arsenic, the unit risk factor is based on EPA's analysis of five data sets of the latest smelter worker epidemiological data collected by four researchers at two smelters (Table 1). To establish a single point estimate, EPA obtained the geometric mean for the data sets within distinct exposed populations and took the final estimate to be the geometric mean of those values. Based on this analysis, EPA used a 0.00429 $\mu g/m^3$ unit risk factor in assessing the health impact of inorganic arsenic.

TABLE 1 Combined Unit Risk Estimates for Absolute-Risk Linear Models

Exposure Source	Study	Unit Risk	Geometric Mean Unit Risk	Final Estimated Unit Risk for Both Smelters
Anaconda smelter	Brown & Chu	1.25×10^{-3}	2.56×10^{-3}	
	Lee Feldstein	2.80×10^{-3}		
	Higgins et al.	4.90×10^{-3}		4.29×10^{-3}
ASARCO smelter	Enterline & Marsh	6.81×10^{-3}	7.19×10^{-3}	
		7.60×10^{-3}		

SOURCE: U.S. Environmental Protection Agency (1986a).

Estimated Public Exposure

EPA applied its Human Exposure Model (HEM) to the 14 smelters to produce quantitative expressions of public exposure to ambient air concentrations. In addition, EPA carried out more site-specific dispersion modeling at El Paso, Texas, and Douglas, Arizona, to evaluate the effects of terrain and buoyancy at fugitive emissions on airborne arsenic concentrations.

Table 2 lists, on a plant-by-plant basis, the total number of people included in the exposure analysis. "Any risk" is the number of people exposed to emissions from the specified source, as calculated by HEM. "Maximum risk" is the number of people exposed to the maximum individual risk from the specified source, as calculated by HEM.

Estimated Individual and Population Risks

By combining numerical expressions of public exposure with the unit risk factor, two types of numerical expressions of public cancer risks are produced. The first, called individual risk, relates to the person or persons who are thought to live in the area of highest concentration as estimated by the computer model. Individual risk is expressed as "maximum individual risk." As used here, the word "maximum" does not mean the greatest possible risk of cancer to the public but is based only on the maximum annual average estimated exposure. The second expression of risk, called population risk, is a summation of all the risks to people estimated to be living within the vicinity (usually within 50 kilometers) of a source. The population risk is expressed as incidences of cancer among all of the exposed population after 70 years of exposure; for convenience, it is often divided by 70 and expressed as cancer incidences per year.

TABLE 2 Number of People Exposed to Emissions

Plant	Total Number of People Exposed		Distance (km) from Source
	Any Risk[a]	Maximum Risk[b]	
ASARCO-El Paso	493,000	1	1.0
ASARCO-Hayden	46,800	1	0.3
Kennecott-Hayden	46,800	1	0.3
Kennecott-Hurley	26,300	1	0.3
Kennecott-McGill	7,350	1	0.3
Kennecott-Garfield	810,000	1	5.0
Phelps Dodge-Morenci	25,500	2	2.0
Phelps Dodge-Douglas	31,000	2	0.2
Phelps Doge-Ajo	6,600	6	
Phelps Dodge-Hidalgo	2,560	909	2.4
Copper Range-White Pine	16,900	1	
Magma-San Manuel	211,000	1	0.2
Inspiration-Miami	35,700	1	0.4
Tennessee Copper-Copperhill	164,000	1	0.5

[a] A 50-kilometer radius was used for the analysis.
[b] People exposed within the distance specified in the next column.

SOURCE: U.S. Environmental Protection Agency (1986a).

Table 3 summarizes the maximum individual risk and the annual incidence for baseline and pollution control scenarios. The baseline level of risk is that resulting from the level of emissions after applying in-place controls or those controls that are required to comply with current state or federal regulations but before applying best available technology (BAT) controls. BAT controls would result in additional reductions of emissions by placing secondary hoods on converter operations. The converter control scenario level of risk is that of the remaining risks after installing BAT controls at all 14 plants.

Uncertainties in Risk Characterization

Exposure for an Entire Lifetime

There are several basic assumptions implicit in the exposure methodology: (1) that all exposure occurs at people's residences, (2) that people stay at the same location for 70 years, (3) that the ambient air concentrations and the emissions that cause these concentrations persist for 70 years, and

TABLE 3 Risk Estimates for Primary Copper Smelters

Smelters	Maximum Individual Risk			Annual Incidence (cases per year)		
	Baseline $\times 10^{-4}$	Converter Control $\times 10^{-4a}$	Reduction $\times 10^{-4}$	Baseline	Converter Control[a]	Reduction
ASARCO-El Paso[b]						
(1)	10	8[c]	2	0.38	0.28[c]	0.09
(2)	6	5[c]	1	0.20	0.18[c]	0.02
(3)	10	9[c]	1	0.18	0.16[c]	0.02
ASARCO-Hayden	13	12	1	0.06	0.05	0.01
Kennecott-Garfield (Utah)	0.6	0.6	0	0.14	0.14	0
Kennecott-Hayden	3	0.5	2.5	0.016	0.0054	0.0106
Inspiration-Miami	1.9	1.0	0.9	0.0069	0.0034	0.0035
Phelps Dodge-Douglas						
(1)[d]	12	2	10	0.022	0.0081	0.0139
(2)	0.8	0.7	0.1	0.025	0.013	0.012
Kennecott-McGill	4	0.6	3.4	0.006	0.0015	0.0045
Phelps Dodge-Hidalgo	0.05	0.03	0.02	0.0001	0.0001	0
Phelps Dodge-Morenci	0.8	0.2	0.6	0.0028	0.0009	0.0019
Phelps Dodge-Ajo	2	1.7	0.3	0.0045	0.0038	0.0007
Kennecott-Hurley	1.2	0.5	0.7	0.0008	0.0003	0.0005

Tennessee Chemical-Copperhill	0.6	0.1	0.5	0.003	0.0006	0.0027
Magma-San Manuel	1.6	0.4	1.2	0.0026	0.0017	0.0009
Copper Range-White Pine	1.1	0.15	0.95	0.0004	0.0002	0.0002

NOTE: The baseline level of risk is that level of emissions that occurs after applying in-place controls or controls that are required to comply with current state or federal regulations but before applying best available technology (BAT). The converter control level of risk is the risk remaining after installing BAT.

[a] Control of converter fugitive emissions by a system consisting of a secondary hood with 94 percent collection efficiency and 95 percent control efficiency.

[b] El Paso figures represent secondary arsenic emissions based on (1) an emission factor for uncontrolled converter fugitive emissions of 15 percent of the arsenic contained in the primary converter process gases, and (2) on a 3.75 percent emission factor. These figures are estimated by EPA to represent the upper and lower bounds of uncontrolled converter fugitive emissions at ASARCO-El Paso. (3) Risk estimates calculated using site-specific analyses (ISCLT/Valley model) and 3.75 percent emission factor.

[c] Risk estimates calculated assuming no additional control by the building evacuation system (BES) of emissions escaping the converter secondary hoods. Some control of these emissions by the BES may occur, although the amount of control cannot be determined. To the extent that emissions escaping the converter secondary hoods are controlled by the BES, these risk estimates are overstated.

[d] Risk estimates calculated using site-specific analyses (ISCLT/Valley model).

SOURCE: U.S. Environmental Protection Agency (1986b:27974).

215

(4) that the concentrations are the same inside and outside a residence. In sum, the exposure methodology assumes that individuals are exposed to inorganic arsenic emissions for their entire lives.

Several reviewers of EPA's methodology have questioned these simplifying assumptions, particularly the assumption of 70-year resident immobility. If EPA used what to the reviewers is a more reasonable assumption—for example, a 10-year residency in the area—then the maximum lifetime individual risk would decrease by approximately one order of magnitude. This 10-year assumption, however, would not change the annual cancer incidence because this calculation is independent of population mobility.

Early Lifetime Exposure

Although the estimates derived from the various epidemiological studies are quite consistent, there are a number of uncertainties associated with them. The estimates were made from occupational studies that involved exposures only after employment age was reached. In estimating risks from environmental exposures throughout life, EPA (1984) assumed in the linear nonthreshold model that the increase in the age-specific mortality rates of lung cancer was a function only of cumulative exposures, irrespective of how the exposure had been accumulated. Although this assumption adequately describes all of the data, it may be in error when applied to exposures that begin very early in life. Similarly, it is possible that linear models are inaccurate at low exposures, even though they reasonably describe the epidemiological data.

Given greater access to the data from these studies, other dose measures, as well as models other than the linear nonthreshold model, could be studied. Such analyses would indicate whether other approaches are more appropriate than the ones applied here.

Use of Census Data

The official EPA risk assessment (Table 3) underestimated the maximum individual risk and cancer incidence at the El Paso and Douglas smelters because it did not include the local Mexican population living in border towns and illegally in the United States. In the case of the El Paso smelter, the population in neighboring Juarez is approximately the same as in El Paso. The maximum individual risk there is similar, 10^{-3}, and the annual cancer incidence ranges from 0.40 to 0.70, which is double the incidence among the U.S. population. These estimates do not include the Mexicans living illegally in the United States because they are not counted by the Census Bureau. In the case of the Douglas smelter, the population of neighboring Agua Prieta is about double the population of Douglas. The maximum individual risk is similar, 10^{-3}, and the annual cancer incidence is 0.04, which is double the incidence among the U.S. population.

If this information were incorporated in risk management decisions, it would lower the cost per life saved and increase the economic efficiency of the regulation, particularly at El Paso.

Assumption of No Latency Period

EPA's risk assessment assumes that there is no latency period between exposure and incidence. Although there is no definitive information on the exact length of the latency period for airborne arsenic, it is greater than zero. Enterline and Marsh (1982) suggest that it may be in the range of 10–19 years because their standardized mortality ratios appear to become significant about 10–19 years after exposed workers have left the plant.

If this information were incorporated in risk management decisions, it would decrease the economic efficiency of the regulation, particularly at El Paso.

Exclusion of Other Health Effects

The unit risk factor used in this case study applies only to lung cancer. Other health effects are possible, however, including skin cancer, hyperkeratosis, peripheral neuropathy, growth retardation and brain dysfunction among children, and increases in adverse birth outcomes. No numerical expressions of risk relevant to these health effects were included in the EPA regulatory analysis.

Evaluation of Risk Assessment

In preparing a risk assessment as a result of reviewing EPA's risk estimates and the uncertainties in the estimates, the analyst or decision maker should focus primarily on the baseline maximum individual risk and annual incidence and assume that the proposed converter controls will achieve the EPA estimated reduction in fugitive emissions. If the baseline level of risk is changed, however, the risk remaining after the installation of converter controls should also be changed proportionally.

RISK MANAGEMENT: EXAMINING THE CONSEQUENCES

After determining the cancer risks from arsenic emissions, the regulatory decision maker should examine the consequences of requiring BAT controls at the 14 copper smelters. This examination could include the estimated emission and risk reduction, the remaining risks after BAT control, the costs and economic impacts, cost-effectiveness and economic efficiency, and the remaining public exposure or equity considerations.

A standard should be determined for emissions from smelter converters that would require all smelters above a specific arsenic feed rate to install pollution controls (column 3 in Table 4). If the standard were 100 kilograms per hour (kg/h) or greater for converter operations, no controls would be required; if it were 0.5 kg/h, controls would be required on converter operations at all 14 smelters.

Emissions and Risk Reductions

The BAT controls would require the installation of secondary hoods on converters. The potential emission reductions from and the estimated annualized costs of the converter secondary controls at each of the existing smelters appear in Table 4. The estimated cost-effectiveness ranges from about $100,000 to $9.7 million per megagram ($/Mg) at the 14 smelters.

Applying BAT controls for converter secondary emissions would reduce the range of estimated maximum individual risks from between 1.0×10^{-3} at the Phelps Dodge-Douglas and 2.0×10^{-5} at the Phelps Dodge-Hidalgo smelters (see Table 3). It would also reduce the estimated annual incidence of lung cancer from between 0.09 (regulating only the plant with the highest incidence, that is, (ASARCO-El Paso) to 0.14 (regulating all 14 plants).

Remaining Exposure and Risks

The remaining exposure and consequent risk are a function of the number of plants that are required to control converter emissions. Applying controls for these emissions at all 14 smelters would change the range of estimated maximum individual risk from between 1.3×10^{-3} and 5.0×10^{-6} to a range of 1.2×10^{-3} and 3.0×10^{-6} (see Table 3). Applying controls would also reduce the estimated annual incidence of lung cancer from a range of 0.38–0.001 to a range of 0.29–0.0001 Applying controls at none of the plants would leave the remaining risks at the same level as the baseline risks.

Costs and Economic Impacts

The annualized control costs per smelter range from a low of $379,000 at ASARCO-El Paso to $3,432,000 at Phelps Dodge-Morenci. If BAT controls were applied to all 14 smelters, the total annualized control costs would be approximately $29 million.

EPA studied the economic impact of imposing controls on the eight smelters with an arsenic feed rate greater than 1 kg/h. For two of the eight smelters, Kennecott-Hayden and Kennecott-McGill, the control costs were likely to result in permanent closure of the smelters. The analysis also indicated that the economic impact would be minimal only at ASARCO-El Paso. An additional factor considered for this smelter was that secondary hoods were scheduled to be installed on all converters to comply with requirements in the Texas state implementation plan for attaining the national ambient air quality standard for lead.

Economic Cost-Effectiveness

To determine whether the standard is cost-effective in terms of number of cancer cases avoided, the analyst must establish a reasonable value for a statistical life saved. The agency's *Guidelines for Regulatory Impact Analyses* (U.S. Environmental Protection Agency, 1983b) suggest that this value should fall in the range of $400,000 to $7 million, with a point estimate of $2 million. This value is supported in part by a recent survey of 130 decisions made by the U.S. government to regulate carcinogens (Travis et al., 1987). The survey found that the average implicit value of a statistical life saved was approximately $2 million.

If $2 million is a reasonable value to consider in a decision to control arsenic, then the figures in Table 5 suggest that not regulating any plant is cost-effective. They show that the cost per case avoided exceeds $2 million at all plants.

If the lowest value, $4.2 million at ASARCO-EL Paso, were considered to be a reasonable value to initiate action, a argument could still be made against controlling fugitive emissions from the converter at that plant. An alternative risk assessment scenario at ASARCO-El Paso, described in Table 3, shows a smaller reduction in cancer incidence (0.02) and consequently a higher value per case avoided ($18.9 million). These figures exceed the highest value in the *Guidelines* and the average value implied by past U.S. government regulatory decisions. The scenario assumes that only 3.75 percent, rather than 15 percent, of arsenic emissions escape the primary vent hood. The higher emission factor of 15 percent reflects actual emissions before El Paso upgraded its gas management system. The lower emission factor of 3.75 percent is an EPA estimate that was derived from the performance of other hoods for which EPA had data rather than from actual measured values at El Paso after installing the new equipment. The new emission factor is probably lower than 15 percent, which means that the cost per incidence avoided is higher than the lower bound of $4.2 million.

TABLE 4 Environmental and Cost Impacts Associated with Secondary Inorganic Arsenic Emission Control Systems for Converter Operations

Smelter	Arsenic Content of Feed (%)	Arsenic Feed Rate to Converters (kg/h)	Potential Secondary Arsenic Emissions (Mg/yr)	Baseline Secondary Arsenic Emissions (Mg/yr)	Predicted Secondary Arsenic Emission Reduction[a] (Mg/yr)	Annualized Control Costs ($000s)	Cost per Unit Emission Reduction ($/Mg)
ASARCO-ElPaso[b]							
(1)	0.5	98.9	98.3	13.3	3.7	379	102,430
(2)	0.5	98.9	24.6	3.4	1.0	379	379,000
ASARCO-Hayden	0.42	63.4	10.2	5.4	4.4	798	181,365
Kennecott-McGill	0.033	9.3	10.1	10.1	9.2	2,201	239,240
Kennecott-Hayden	0.015	7.2	6.5	6.5	5.9	2,140	362,710
Phelps Dodge-Douglas	0.03	4.2	4.1	4.1	3.7	2,943	795,405

Inspiration-Miami	0.033	5.7	1.9	1.9	1.7	2,943	1,731,000
Phelps Dodge-Morenci	0.006	1.9	1.9	1.9	1.7	3,432	2,019,000
Kennecott-Utah (Garfield)	0.144	14.7	1.5	1.5	1.4	2,028	1,449,000
Phelps Dodge-Hidalgo	0.003	0.4	0.2	0.2	0.18	1,745	9,694,000
Tennessee Chemical-Copperhill	0.0004	0.7	0.65	0.65	0.58	1,278	2,203,000
Magma-San Manuel	0.006	0.6	0.55	0.55	0.50	3,979	7,958,000
Phelps Dodge-Ajo	0.015	0.8	0.52	0.52	0.47	1,562	3,323,000
Kennecott-Hurley	0.0005	0.8	0.46	0.46	0.42	2,296	5,467,000
Copper Range-White Pine	0.008	0.5	0.30	0.30	0.27	1,278	4,733,000

[a]Emission reduction estimates calculated, assuming no additional control by the building evacuation system (BES) of emissions escaping the converter secondary hoods. Some control of these emissions by the BES may occur, although the amount of control cannot be determined. To the extent that emissions escaping the converter secondary hoods are controlled by the BES, these emission reductions are understated.

[b]El Paso figures represent secondary arsenic emissions based on (1) an emission factor for uncontrolled converter fugitive emissions of 10 percent of the arsenic contained in the primary converter process gases and (2) a 3.75 percent emission factor. These figures are estimated by EPA to represent the upper and lower bounds of uncontrolled converter fugitive emissions at ASARCO-El Paso.

SOURCE: U.S. Environmental Protection Agency (1986b:27973).

TABLE 5 Annual Cost per Case Avoided and Net Present Value (NPV) of Controls (in millions of 1982 dollars)

Plant	Annual No. of Reduced Cases	Annual Cost/Case (10%, 0 yrs)	NPV[a] (Ben-Cost) (4%, 0 yrs)	NPV[a] (Ben-Cost) (10%, 0 yrs)	NPV[a] (Ben-Cost) (4%, 15 yrs)	NPV[a] (Ben-Cost) (10%, 15 yrs)
ASARCO-El Paso						
(1)	0.09	4.2	-1.6	-1.7	-2.7	-2.9
(2)b	0.02	18.9	-3.5	-2.9	-3.7	-3.1
(3)	0.02	19.0	-3.5	-2.9	-3.7	-3.1
ASARCO-Hayden	0.01	79.8	-8.4	-6.6	-8.5	-6.8
Kennecott-Garfield	0.	n.a.	-22.3	-17.3	-22.3	-17.3
Kennecott-Hayden	0.0106	201.9	-24.0	-18.0	-24.2	-18.2
Inspiration-Miami	0.0035	840.9	-36.5	-25.0	-36.6	-25.0
Phelps Dodge-Douglas						
(1)c	0.0139	185.1	-36.2	-24.8	-36.4	-25.0
(2)c	0.012	245.2	-36.3	-24.9	-36.4	-25.0
Kennecott-McGill	0.0045	488.1	-25.5	-18.7	-25.6	-18.7
Phelps Dodge-Hidalgo	0.	n.a.	-21.6	-14.9	-21.6	-14.9
Phelps Dodge-Morenci	0.0019	1,806.3	-38.9	-29.2	-38.9	-29.2
Phelps Dodge-Ajo	0.0007	2,231.4	-19.1	-13.3	-19.1	-13.3
Kennecott-Hurley	0.0005	4,592.0	-28.4	-19.5	-28.4	-19.5
Tennessee Copper-Copperhill	0.0027	473.3	-15.9	-10.8	-15.9	-10.9
Magma-San Manuel	0.0009	4,421.1	-49.8	-33.9	-49.8	-33.9
Copper Range-White Pine	0.0002	6,390.0	-16.0	-10.9	-16.0	-10.9

[a]The benefit estimate is based on $2 million per statistical life.
[b]Risk estimates calculated using site-specific analyses (ISCLT/Valley model) and 3.75 percent emission factor.
[c]Risk estimates calculated using site-specific analyses (ISCLT/Valley model).

SOURCES: Derived from Tables 3 and 4.

222

One could dispute the claim about the unreasonableness of the cost per case avoided on the grounds that the estimate fails to include the benefits to the Mexican population. Including the Mexican population near the El Paso smelter and using a high emission rate reduce the annual incidence by 0.265, as opposed to the 0.09 reduction achieved with controls on fugitive emissions. Consequently, the cost per case avoided is $1.4 million, rather than $4.2 million. Including the Mexican population and using a low emission rate reduce the annual incidence by 0.12, compared with the 0.2 reduction achieved with controls on fugitive emissions. In this instance, the cost per case avoided is $3.2 million, rather than $18.9 million. These revised estimates suggest that the cost per case avoided at El Paso is consistent with EPA *Guidelines* and other U.S. government decisions. However, including the Mexican population in the estimates for the Douglas smelter only brings the cost per case avoided nearer to the high end of the range in the EPA *Guidelines*.

Economic Efficiency

One can determine whether a standard is economically efficient by comparing the net present value of the standard's benefits and costs (Table 5). As the table shows, with a 10 percent discount rate and no latency period, no control option has a positive net present value (assuming $2 million per statistical life saved). Changing the discount to 4 percent does not significantly change the results. The net present value of all controls is negative. Thus, regulation at all plants would be rejected on the grounds of economic efficiency.

Using a 15-year latency period changes the economic efficiency evaluation only at the El Paso smelter. The magnitude of the change in the negative net present value is greater for this plant because of the greater proportionate reduction in the values for cancer cases avoided. The assumption of a 15-year latency does not markedly alter the negative economic efficiency evaluation at the other 13 plants because fewer cases are avoided and because the control costs overwhelm the benefits in the net present value calculations.

Assuming a value of $7 million per statistical life saved only partly alters the negative economic efficiency valuation at the El Paso smelter. Using the highest reduction in annual cancer incidence—0.09 cases—makes the net present value positive for both discount rates with no latency period and for the 4 percent discount rate with a 15-year latency period (not shown in Table 5). In all other circumstances at El Paso (0.09 incidence/10 percent/15 years and 0.02 incidence/any combination of rates and years) and at all other smelters, the net present value would remain negative, even for a relatively high value per statistical life saved.

TABLE 6 Health and Welfare Benefits from Reducing Lead and Particulate Matter Emissions

Pollutant Effect	Number of Reduced Effects/Ton of Emission Reduction[a]	Benefits ($/ton) of Emission Reduction
Lead		
Chelation	4.5 fewer cases	16.5×10^3
Hypertension	47.0 fewer cases	10.5×10^3
Myocardial infarction	0.14 fewer cases	8.8×10^3
Strokes	0.03 fewer cases	1.4×10^3
Mortality	0.13 fewer deaths	260.0×10^3
Total for lead		297.2×10^3
Particulate Matter		
Mortality	4.3×10^{-5} fewer deaths	0.1×10^3
Morbidity		
Lost workdays	8.8 fewer sick work loss days	0.7×10^3
Reduced-activity days	27.9 fewer sick work loss hours	1.1×10^3
Medical expenditures	--	0.3×10^3
Soiling and material damages	--	0.2×10^3
Total for particulate matter (national average)		2.4×10^3

[a]Effects based on county-weighted average for estimates of reduced number of effects per ton of reduced pollution emissions. Range for the particulate matter total benefit value varies between $0 and $300,000/ton for individual counties in the United States. The variation by county for the benefits of reducing lead emissions was not calculated.

SOURCE: U.S. Environmental Protection Agency (1983a, 1985).

In making a regulatory decision, one could also examine whether the standard would be economically efficient if the total benefits of emission reductions were compared with the costs. At El Paso, installation of BAT controls for fugitive arsenic from converters would also achieve a 6.6-Mg reduction in lead (Pb) and a 30-Mg reduction in particulate matter (PM). Controls at Douglas would reduce PM by 780 Mg. An immediate issue, however, is how to compare and aggregate these health and welfare benefits. In the case of PM, there are reductions in mortality, morbidity, and welfare damage to be considered (Table 6). EPA's *Guidelines* encourage the analyst to monetize these benefits and then add them together to obtain a dollars-per-ton figure.

The PM benefit/ton typically ranges from $300/ton to $10,000/ton, depending on the PM sources and exposure patterns for receptors near the sources. Incorporating the PM benefits into an economic analysis would not alter the efficiency evaluation at the Douglas plant if one accepted

TABLE 7 Net Present Discount Values for Benefit-Cost Streams of Particulate Matter (PM), Lead, and Arsenic Emission Reductions

Plant/Pollutant	Reduced Emissions (tons)	Present Value Benefits[a] ($ million) Discount Rates		Net Present Value ($ million) Discount Rates	
		4%	10%	4%	10%
ASARCO-El Paso					
Lead	6.6	26.7	16.7		
PM	29.5	1.2	0.8		
Arsenic[b]					
Scenario 1	3.7	2.4	1.5		
Scenario 2	1.0	0.5	0.3		
Total[c]					
Scenario 1		30.3	19.0	26.3	15.8
Scenario 2		28.4	17.8	24.4	14.6
Phelps Dodge-Douglas					
PM	776	10.5	6.6		
Arsenic[d]					
Scenario 1	3.7	0.4	0.3		
Scenario 2	3.7	0.3	0.2		
Total[e]					
Scenario 1		10.9	6.9	-25.7	-18.2
Scenario 2		10.8	6.8	-25.7	-18.2
Scenario A				> 0.0	
Scenario B					> 0.0

[a]Assumes no latency period for assessing health and welfare effects for PM, lead, and arsenic.
[b]ASARCO-El Paso Scenario 1 assumes that 15 percent of the arsenic emissions escape the primary vent hood; Scenario 2 assumes that 3.75 percent of the arsenic emissions escape.
[c]The expected benefit/ton value for reductions in PM is $3,000/ton for the El Paso area.
[d]Phelps Dodge Scenario 1 risks are based on a standard exposure analysis. Scenario 2 risks are based on site-specific analysis (ISCLT/Valley model).
[e]The expected benefit/ton value for reductions in PM is $1,000/ton for the Douglas area. To acheive a net present value greater than zero with a discount rate of 4 percent (Scenario A), the PM benefit/ton value must be at least $3,000/ton. The benefit/ton value given a 10 percent discount rate (Scenario B) must be at least $3,800/ton.

the population weighted value for the Douglas area of $1,000/ton (Table 7). However, using a value of $3,800/ton makes the benefits equal to the costs, in contrast to the negative net present value of $25–$36 million for which no PM benefits are considered, or a negative $18–$26 million for the scenario with $1,000/ton PM benefits.

Incorporating the PM benefits into an economic analysis of controls at the El Paso plant does not alter the negative efficiency calculation, given the population weighted value for the El Paso area of $3,000/ton, nor any value within the range of $300–$10,000/ton. Incorporating the lead benefits, however, yields a positive net present value of $15–$26 million, rather than a negative net present value of $0.4–$2.3 million.

Equity

The regulatory decision maker may also consider whether a standard meets an equity objective rather than an economic efficiency objective. For example, it might be decided that controls should apply in all circumstances in which maximum individual risk equals or exceeds 10^{-4}. In 1986, this risk level existed at three smelters: El Paso, Hayden, and Douglas. Alternatively, controls might be applied to the degree necessary to reduce all risks to at least 10^{-5}. This goal is not technically possible with the proposed BAT controls, however, because maximum individual risk remains greater than 10^{-5} even after the installation of BAT controls at all plants.

FACTORS TO CONSIDER IN SETTING A STANDARD

The risk assessment and risk management information presented in the two previous sections is all that is available to make a regulatory decision. As the regulatory decision maker, you the reader must decide how many of the 14 plants to regulate by expressing a standard that imposes converter control hoods on plants above a specific arsenic feed rate. The following points should receive careful consideration.

How Should Health Risk Be Characterized?

The quantitative information about health risk applies only to lung cancer. How would you account for other possible health effects (described in the second section)?

The quantitative estimate of lung cancer risk appears to be both an overstatement and understatement of risk, as described earlier. It is an overstatement because it assumes continuous exposure for 70 years and does not take into account the potential latency period between exposure and incidence. It is an understatement because it excludes the exposed Mexican population. (As a practical matter, it should be noted that the Clean Air Act does not authorize extraterritorial jurisdiction and thus cannot be applied to Mexico.) It could be further over- or underestimated depending on how one deals with the uncertainties in the epidemiological studies.

What Constitutes a Significant Risk?

In determining what constitutes a significant risk, you should consider both maximum individual risks and annual cancer incidences resulting from exposure to inorganic arsenic. The current maximum individual risks range from 1.3×10^{-3} to 5.0×10^{-6}; the annual incidences range from 0.38 to 0.0001. At some smelters, both the individual and population risks are low and would probably be deemed insignificant. In your decision making, you should determine the level of baseline risk you think is insignificant and thus would not warrant regulation.

Your decision is complicated by the lack of a perfect correspondence among plants on individual and population risks. For example, the individual risk exceeds 10^{-4} at three plants. Yet, the baseline cancer incidence varies by more than an order of magnitude among the three plants: the highest incidence, at El Paso, is 0.38; the lowest incidence, at Douglas, is 0.022. Consequently, although the significance of the risk generated by the plants is the same for individual risk, it is very different for annual incidence.

What Constitutes an Appropriate Balance Between Costs and Risks?

One approach that you might consider in addressing this issue is an economic cost-effectiveness cutoff for incidence (cases of disease) avoided. As noted in the last section, the cost per incidence avoided ranges from $4.2 million to $6.39 billion. Is it reasonable to exclude some plants from regulation on the basis of this information?

Also, as described in the last section, the cost per incidence avoided at two plants (El Paso and Douglas) changes from the official EPA analysis if it includes the exposed Mexican population. The cost per incidence avoided ranges between $1.4 million and $3.2 million rather than between $4.2 million and $18.9 million at El Paso if the Mexican population is included; the cost per incidence avoided still exceeds $7 million at Douglas if the Mexican population is included in the analysis. Does inclusion of the information change what you think about the necessity for regulation?

Another approach that you might consider in addressing the issue of balance between costs and risks is the economic efficiency of the standards. Under most conventional approaches, BAT controls at all plants are economically inefficient because the costs exceed the benefits. Even the use of the preferred EPA approach, which assumes no latency period between exposure and incidence and a low discount rate, does not alter the conclusion that the net present value of controlling any plant is negative.

Nevertheless, including the "cocontrol" benefits of reduced lead and particulate matter together with the arsenic benefits alters these calculations at one plant. Incorporating all pollution reduction benefits at the El Paso and Douglas smelters suggests that the imposition of BAT controls is economically efficient at the El Paso plant but is still not efficient at the Douglas plant.

How Should Single-Decision Criteria Be Explicitly Integrated?

You have received information about several decision criteria to assist you in making a regulatory decision. These criteria include estimates of maximum individual risk, annual incidence of cancer, cost per life saved, economic efficiency of controls, and economic impacts on the copper smelter industry, as well as the uncertainties in these estimates. In addition, you know that for the El Paso smelter the state implementation plan would result in a significant reduction of inorganic arsenic emissions even if EPA did not issue a standard for inorganic arsenic. Also, you know that there would be, as described in the previous section, cocontrol of other pollutants with a standard for inorganic arsenic. Can you be explicit about the relative importance of these single-decision criteria and how you combined them, together with the qualitative factors, to arrive at your decision to impose controls at specific plants?

Is Any Balance Between Costs and Risks Consistent with EPA's Legislative Mandate?

An overriding risk management issue is whether consideration of economic information is consistent with the requirements of the Clean Air Act. As required by Section 112 (b)(1)B of the act, standards must be set "at the level which in [the administrator's] judgment provides an ample margin of safety to protect the public health" from inorganic arsenic emissions.

EPA interprets Section 112 to require a judgment about the degree of control that can be considered amply protective. For nonthreshold pollutants, two choices are available: (1) to set the standards at zero emissions to eliminate the attributable health risks, or (2) to permit some residual risk. In the absence of a specific directive on this choice in Section 112, and in recognition of the drastic economic consequences that could follow from a requirement to eliminate all risk from hazardous air pollutant emissions, EPA believes that it is not the intent of Section 112 to eliminate all risks. Standards that permit some level of residual risk can be considered to provide an ample margin of safety to protect public health. Therefore, EPA maintains that there must be a consideration of costs in regulating hazardous air pollutants.

The Natural Resources Defense Council (NRDC) is currently contesting EPA's interpretation of Section 112. It argues that EPA may not consider nonhealth issues, such as technology, economics, and affordability. Furthermore, NRDC holds that public health should be the sole consideration in developing Section 112 standards, with no consideration of such factors as cost and the availability of technology.

You have an opportunity to address this broader issue. If the EPA position is correct, what information should EPA consider in deciding on a level of pollution control that may present some human health risk? The EPA position in the arsenic rule making was that the administrator "selects a level of control which, in his judgment, reduces the health risks to the greatest extent that can reasonably be expected, after considering the uncertainties in the analysis, the residual risks remaining after the application of the selected control level, the costs of further control, and the societal and other environmental impacts of the regulation" (U.S. Environmental Protection Agency, 1986b:27958). The EPA position excludes consideration of economic cost-effectiveness and economic efficiency. If the NRDC position is correct, EPA ought not to consider any economic factors. In this situation, EPA should impose BAT controls on all 14 smelters at an annual cost of $29 million to prevent 0.134 annual cancer incidence. The cost per incidence avoided would be $216 million, a figure that does not suggest a reasonable balance between costs and risk reduction.

EPA'S ACTUAL REGULATORY DECISION FOR INORGANIC ARSENIC EMISSIONS FROM PRIMARY COPPER SMELTERS

On June 5, 1980, EPA published a *Federal Register* notice (U.S. Environmental Protection Agency, 1980) listing inorganic arsenic as a hazardous air pollutant under Section 112 of the Clean Air Act.

On July 11, 1983, EPA proposed standards (U.S. Environmental Protection Agency, 1983c) for inorganic arsenic emissions from the nation's 14 low-arsenic primary copper smelters, as well as for high-arsenic copper smelters and glass manufacturing plants. The proposed standard for low-arsenic primary copper smelters regulated secondary inorganic arsenic emissions from converter operations and from matte and slag-tapping furnaces. The proposed standards for converter operations applied to smelters with an annual average inorganic arsenic feed rate of 6.5 kg/h or greater. The proposed standards for matte and slag-tapping furnaces applied to smelters with an annual average combined inorganic arsenic process rate of 40 kg/h or greater. (The latter standards were less restrictive than the former because secondary emissions from coverters are typically 1 to 25 times greater than the combined emissions from both matte and slag-tapping operations.)

The proposed standards affected 6 of the existing 14 low-arsenic primary copper smelters. The estimated capital and annualized costs required to meet the standards were approximately $35.3 million and $9.5 million, respectively.

Following public comment on the proposed standards, EPA conducted additional analyses to ensure that the final rule was based on the most complete and accurate information available. These additional anlyses included revising the emission estimates, the exposure concentration estimates, and the risk assessment, as well as conducting additional cost and economic impact analyses. The scope of these analyses resulted in considerable changes in the risk assessment and risk management information that was incorporated in the final rule making of August 4, 1986 (U.S. Environmental Protection Agency, 1986b).

Using the revised risk and cost estimates, EPA concluded that for eight copper smelters (Inspiration-Miami; Phelps Dodge-Hidalgo, -Morenci, and -Ajo; Kennecott-Hurley; Tennessee Chemical-Copperhill; Magma-San Manuel; and Copper Range-White Pine), the baseline risk was less than or equal to the risks of the standards it had previously withdrawn and that regulation was not warranted. For five of the six remaining smelters (all but ASARCO-El Paso) that were affected by the proposed standard, EPA concluded that the costs were disproportionate to the risk reductions that could be obtained. Furthermore, the economic analysis showed that for two of these five smelters (Kennecott-Hayden and Kennecott-McGill), the control costs were likely to result in the smelters remaining permanently closed.

For the remaining facility, ASARCO-El Paso, the analysis indicated that risk could be reduced at a reasonable cost. An additional factor considered in the assessment was that secondary hoods were to be installed on all converters at ASARCO-El Paso to comply with requirements in the Texas state implementation plan for attainment of the national ambient air quality standard for lead. Because the costs of control in this instance are reasonable and the controls can be implemented now, EPA decided that these controls should be applied only at ASARCO-El Paso. As a result, the final standard affects only converter operations at 1 of the 14 smelters and does not apply to matte and slag-tapping operations. It applies only to smelters with annual arsenic feed rates to converters of greater than 75 kg/h. The estimated capital and annualized costs required to meet the final standard are approximately $1.8 million and $380,000, respectively.

Because the only high-arsenic copper smelter affected by the earlier proposal ceased operation in 1985 (a plant owned and operated by ASARCO in Tacoma, Washington), EPA withheld further action on the proposed standard for high-arsenic primary copper smelters.

REFERENCES

Brown, C.C., and K.C. Chu
  1983   Implications of the multi-stage theory of carcinogenesis applied to occupational arsenic exposure. *JCNI* 70:455-463.
Enterline, P.E. and G.M. Marsh
  1982   Cancer among workers exposed to arsenic and other substances in a copper smelter. *American Journal of Epidemiology* 116:895-911.
Higgins, I.T.T., K.B. Welch, M.S. Oh, et al.
  1985   Arsenic Exposure and Respiratory Cancer in a Cohort of 8,044 Anaconda Smelter Workers. Report to the Chemical Manufacturers Association and the Smelter Environmental Research Association. October.
Lee-Feldstein, A.
  1986   Cumulative exposure to arsenic and its relationship to respiratory cancer among copper smelter employees. Journal of Occupational Medicine 38:196-302.
Travis, C.C., S.A. Richter, E.A.C. Crouch, R. Wilson, and E. Klema
  1987   Cancer risk management by federal agencies. *Environmental Science and Technology* 21:415-420.
U.S. Environmental Protection Agency
  1980   National emission standards for hazardous air pollutants: Addition of inorganic arsenic to list of hazardous pollutants. *Federal Register* 45(110):37886-37888.
  1983a   *Benefits and Net Benefits Analysis for Alternative National Ambient Air Quality Standards for Particulate Matter*. Final Regulatory Impact Analysis. Washington, D.C.: Environmental Protection Agency, Office of Air Quality Planning and Standards.
  1983b   *Guidelines for Performing Regulatory Impact Analysis*. EPA-230-1-84-003. Washington, D.C.: Environmental Protection Agency.
  1983c   National emission standards for hazardous air pollutants: Proposed standards for inorganic arsenic. *Federal Register* 48(140):33112-33180.
  1984   *Health Assessment Document for Inorganic Arsenic*. EPA-600/8-83-021f.
  1985   *Costs and Benefits of Reducing Lead in Gasoline*. Final Regulatory Impact Analysis. EPA-230-05-85-006. Washington, D.C.: Environmental Protection Agency, Office of Policy Analysis.
  1986a   *Inorganic Arsenic Emissions from Primary Copper Smelters and Arsenic Plants—Background Information for Promulgated Standards*. EPA-450/3-83-010b. Washington, D.C.: Environmental Protection Agency.
  1986b   National emission standards for hazardous air pollutants: Standards for inorganic arsenic. *Federal Register* 51(149):27957-28042.